Everyday Moral Economies

RGS-IBG Book Series
Published

Forthcoming

Everyday Moral Economies

Food, Politics and Scale in Cuba

Marisa Wilson

Chancellor's Fellow at the School of Geosciences
University of Edinburgh

WILEY Blackwell

This edition first published 2014
© 2014 John Wiley & Sons, Ltd

Registered Office

John Wiley & Sons, Ltd, The Atrium, Southern Gate, Chichester, West Sussex, PO19 8SQ, UK

Editorial Offices

350 Main Street, Malden, MA 02148-5020, USA

9600 Garsington Road, Oxford, OX4 2DQ, UK

The Atrium, Southern Gate, Chichester, West Sussex, PO19 8SQ, UK

For details of our global editorial offices, for customer services, and for information about
how to apply for permission to reuse the copyright material in this book please see our
website at www.wiley.com/wiley-blackwell.

Library of Congress Cataloging-in-Publication Data

Wilson, Marisa L. (Marisa Lauren), 1979–
  Everyday moral economies: food, politics and scale in Cuba / Marisa Wilson.
    pages   cm – (RGS-IBG book series)
  Includes index.
  ISBN 978-1-118-30200-2 (hardback) – ISBN 978-1-118-30192-0 (paper)
1. Food supply – Social aspects – Cuba.   2. Food supply – Economic aspects – Cuba.
3. Consumption (Economics) – Cuba.   4. Exchange – Cuba.   5. Value.   6. Cuba –
Economic conditions – 1990–.   I. Everyday moral economies.
  HD9014.C92W55 2014
  338.1'97291—dc23

                                    2013018233

A catalogue record for this book is available from the British Library.

Cover image: 'The hand is that of a woman farmer in her mid-60s, who requested that the
photo be taken as a symbol of "a real worker in Cuba". As she told me, "you can always tell
a *campesino* [farmer] by their hands".' © Marisa Wilson
Cover design by Workhaus.

Set in 10/12pt Plantin by SPi Publisher Services, Pondicherry, India
Printed in Malaysia by Ho Printing (M) Sdn Bhd

1   2014

For my parents

Contents

Series Editors' Preface

The RGS-IBG Book Series only publishes work of the highest international standing. Its emphasis is on distinctive new developments in human and physical geography, although it is also open to contributions from cognate disciplines whose interests overlap with those of geographers. The Series places strong emphasis on theoretically informed and empirically strong texts. Reflecting the vibrant and diverse theoretical and empirical agendas that characterize the contemporary discipline, contributions are expected to inform, challenge and stimulate the reader. Overall, the RGS-IBG Book Series seeks to promote scholarly publications that leave an intellectual mark and change the way readers think about particular issues, methods or theories.

For details on how to submit a proposal please visit:
www.rgsbookseries.com

Neil Coe
National University of Singapore

Joanna Bullard
Loughborough University, UK
RGS-IBG Book Series Editors

Preface

¡Con lo que un yanqui ha gastado
no más que en comprar botellas
se hubiera Juana curado! ...
With what a Yankee spends
Just buying bottles,
Juana could have been cured! ...
                    Nicolas Guillén (from the poem,
                    Visita á un solar, 1930)[1]

This book is about the relationship between provisioning and politics. To be clear, politics is understood in terms of values, economic or otherwise. In this sense, politics is 'less about the struggle to appropriate value (or freedom to create/accumulate value), but the struggle to establish what value *is* (or the freedom to decide what makes life worth living)' (Graeber 2001: 88). I am concerned with values and their spatio-temporal dimensions, like nationalism or economic globalization, and with the way associated values are evidenced in moral ideas and practices that shape everyday life.

In the above verses, for example, there are two values of beer: the first is the market value paid for by tourists from the United States, the second, the social value of finding a cure for Juana (a poor woman from rural Cuba). As the poem suggests, in the 1930s ordinary Cubans saw the two forms of value as commensurable; 'Yankees' did not. Since then, contradictions between social values and market values have become even more pronounced, associated with incessant bi-polar discourses of liberalism and socialism. As I will argue, each discourse is tied to particular temporalities and spatialities, becoming what I call *Leviathans*[2] that frame the material and ideational spaces in which ordinary people in Cuba claim their rights and entitlements.

Officially if not always empirically, values set by markets such as price stand in direct contrast to welfare values such as the grand narrative of Cuban socialism, according to which necessities such as food are considered human rights, distinct from the world of commodities. In this normative scheme, basic foodstuffs should be accessible to all needy Cubans in domestic currency, pesos, though more desirable items may only be available in hard currency (or in equivalent peso prices). The traditional planned economy of Cuba is based on a model that treats the nation as one socialist enterprise, whose ultimate aim is not profit (surplus value) but to ensure alimentary and other needs (social values) of the national community. The scalar project of Cuban nationhood, which controls and rationalizes collective forms of provisioning, and the global political economy that gives some Cubans more options than others, are practical effects of these contrasting normative and material systems, the one that privileges the sovereign nation, the other, the sovereign consumer. This book reveals how people in rural Cuba rationalize the practicalities of living in this contradictory moral and political economic world, in which both national and supranational norms influence rather than determine a more localized politics of value-making.

It was this interest in the relation between values and experience, and in the moralities, materialties and spatialities of this relation, that first motivated me to write this book. My own concern with food politics developed when I spent time in Cuba observing and often living through Cubans' daily 'fight' (*lucha*) to provision food for their families. As an ethnographic researcher, my analysis had to start with the 'concrete conditions which stimulate interest in some abstract problems rather than others' (Hart 1986: 637), and so naturally I focused on the main concern of the people under study: food. As someone from a country with much influence over the global political economy of food, the topic of food politics was also personal.

For at least a decade, I have been struck by the historical divergence of values that have developed over time in my 'home' – the United States – from those that emerged in Cuba, a country located just 90 miles away. Growing up in the Central Valley of California, I witnessed the large-scale conversion of prime agricultural lands into residential or commercial properties, creating what geographers call a 'spatial fix' that cannot easily be undone. As I was to discover, an opposite pattern was happening in 1990s Cuba, where prior neglected and/or damaged land was being converted to agroecological production to provide food for Cubans. This shift in land use patterns is a reflection of two different ways that powerful interests in each country have come to value land and its products: the first that sees land as a means to acquire high rents and profit, the second that sees land as a means to ensure collective entitlements. In the case of the United States and most other countries where private agro-food interests have come to overpower (or accord with) public regulation, food is treated primarily as a commodity. In the case of Cuba, food for domestic consumption is officially a public good, though it may also become a commodity in export, tourist and local farmers' markets. While even organic food production in California must yield enough profits to outcompete residential or commercial land developers (see Guthman 2004), Cuban food production is guided more by alimentary necessity than market determinants.

The aims of agroecological food production in Cuba may seem ideal to the radical, ecologically minded westerner. But in the event that they could personally choose between social and market values (and he or she is more likely to have this choice than Cubans), they may not like to leave their preference for a salad of organic basil – ready washed and served with fresh mozzarella, organic heirloom tomatoes and Californian olive oil – for a collective value system that serves a simple salad of peeled cucumbers and soya oil. Indeed, it is all too easy to idealize the Cuban experience as an admirable alternative to our own, forgetting all the privileges of the market that we as 'responsible' consumers take for granted (forgetting too that ordinary Cubans would likely want access to such privileges if given the opportunity). Actually, it is this very dichotomy between 'us' and 'them' – and between market and collective (or state) forms of value – that is problematized in this book.

On a more theoretical level, the book reveals what Neil Smith (1992: 78) calls the 'double-edged nature of scale' as both enabling and disabling different forms of value: 'By setting boundaries, scale can be constructed as a means of constraint and exclusion, a means of imposing identity, but a politics of scale can also become a weapon of expansion and inclusion, a means of enlarging identities' (Smith 1992: 78). While on the one hand Cuba's food politics often limit the value of food to an instrumental substance to satisfy collective needs, neglecting consumer demand and choice, on the other, Cuba's scalar politics of food reinforce long-term values for national sovereignty and social (and now, environmental) justice, which ordinary people elicit in their own definitions of what it means to be Cuban. At the level of everyday experience, where serendipitous events and encounters enable certain forms of value and disable others, people sometimes maintain, sometimes contest what I call Cuba's national moral economy.

In a sense, then, this book is about possibilities. It is about the transformative capacities of ordinary people in rural Cuba who must work within and between internal and external materialities and moralities. It is also about analytical possibilities that emerge when one shifts from western dichotomies – between fixed representations and unfixed flows or networks, for example – to the *creative formation* of such abstract representations as Cuban nationhood, which are, ironically, often the result of unfixed, cross-border interactions.[3] As Marilyn Strathern (1995: 29) argues: 'Abstract knowledge is an end-result, the effect of creative work ... In short, output cannot be measured against input, for they involve activities of different scale'. One way to illuminate such 'creative work' is through ethnographic experience, which reveals, among other things, the political, analytical and *spatial* potentialities of economic life as the constant formation and re-formation of values and relations.

Crossovers in anthropology and geography I

The fact that *we* stand at one of the poles of comparison is not without its use in clarifying the whole comparative set. Maybe this is the main point: we are back to what we call radical comparison, a comparison in which we ourselves are involved. (Dumont 1986 [1983]: 8)

Understanding the potentialities of everyday economic life in Cuba and elsewhere necessitates a relational view of one's own values vis-á-vis those of one's ethnographic 'informants', but also a relational view of such values vis-á-vis different spatial levels. In what follows, I use the term 'level' to differentiate empirical spaces of different size or extent: the individual, the household, the state, the region, the world. By contrast, I use the term 'scale' to refer to normative projects for community defined in terms of a particular level: the *maximizing* individual, the *nation* state, the *globalizing* world.[4]

Though they have generally preferred the terms 'unit' or 'border' to scale, social anthropologists' emphasis on how people become positioned relative to wider forms of identification is not far from recent concerns about the relational qualities of scale and space in human geography (for more on this connection, see 'Crossovers in anthropology and geography II' in chapter 1). One thing that makes geography different from most anthropology is the pursuit of normative conclusions, especially in the context of an increasingly powerful discourse of economic globalization. Conversely, anthropology has not traditionally provided normative arguments of what 'should' happen in the future (Rankin 2004: 3); rather, the anthropologist aims to elucidate the perspective of an ethnographic 'apprentice' (Jenkins 1994) attempting to make the world he or she encounters in the 'present'[5] and, in turn, their own world, evident through writing and reinforced with historical and comparative analysis.[6]

The issue that unites the two disciplines that shape this book is the multiple affiliations people make that cut across different value systems, some more individualized, some more communitarian. These values may be seen as scalar politics in the sense that they are attached to political projects for personal ambition or to wider projects for community. My use of the word 'scalar' is borrowed from the idea in mathematics and physics, meaning 'having only magnitude [or size], not direction' (OED 2008: 918). Thus 'scalar' points to the multiple identifications particular individuals may have which have no predetermined direction. A person may move from an interest in the personal to the familial, national or indeed the transnational scale. Women from the United States who identify with their female compatriots during a presidential election may at another point in time identify with women from South Sudan as part of a wider feminist community (Sen 2009: 137). Alternatively, and according to circumstance, affiliations may move in the opposite direction, from a wider understanding of community (identifying with England during the World Cup) to a more geographically narrow definition of identity (siding with a Premiership team, say, West Ham United), depending on which forms of affiliation are at play in a particular context (Gellner 2010).

The theme of shifting scales of identification has a long history in anthropology, from Sir Edward Evan Evans-Pritchard's (1940) discussion of the segmentation of southern [now South] Sudanese tribes and lineages into domestic, political and ecological groupings, to Louis Dumont's (1980 [1966], 1986 [1983]) illustration of the way modern values like individualism 'encompass' and thus eclipse opposing values evidenced in everyday life, like hierarchy. More recently, social anthropologists like Marilyn Strathern (1995) have complicated simple accounts of scale that are depicted, for example, in borders that separate the person/body from the non-human world, a distinction not relevant to all cultures (e.g. the Dobu of Papua New Guinea; Strathern 1995: 15).

In geography, the realization that there is no one-to-one correspondence between dominant forms of identification and everyday social relations has been dealt with more recently, but perhaps in terms of more pressing matters of the 21st century such as whether and how people in affluent countries maintain partiality for themselves and their families while taking up historical responsibilities that involve the livelihoods of distant others (Massey 2011 [2005]; Gibson-Graham 1996, see also Young 2011). Drawing from political philosophers such as Nancy Fraser (i.e. 1996, 2009) and Iris Marion Young (1990, 2011), recent work in human geography seems to come closest to substantiating the analytical and political importance of relating identity and scale to issues of justice. Among geographers who have written in this vein are Doreen Massey (2011 [2005]), who uncovers the political limitations of singular accounts of 'local' or 'global' forms of identification, and the feminist duo Gibson-Graham (i.e. 1996, 2006), who reveal the empirical inadequacy of a singular 'capitalocentric' economic identity. Both Massey and Gibson-Graham will be important in this book, but so will earlier theorists of value and relation in anthropology like Louis Dumont (1980 [1966], 1977, 1986 [1983]).

Opening up both anthropological and geographical analysis to cut across values and identifications, which have no predetermined direction, counteracts prior errors in social science that centre on a bordered community (e.g. the nation state) as the primary unit of analysis, without taking into account other forms of affiliation: 'As soon as we introduce trans-border interactions, we admit the possibility of multiple non-isomorphic structures, some local, some national, some regional, and some global, which mark out a variety of different "who's" for different issues' (Fraser 2009: 39). Recent work in both geography and anthropology highlights the importance of such multiple and shifting identifications, counteracting earlier social scientific accounts that reified dominant forms of representation like the national economy. But the baby of representation should not be thrown out with the bathwater of reification, for we cannot neglect what I have earlier called the creative formation of (scalar) representations (following Strathern 1995: 29). Indeed, to focus simply on heterogeneous forms of identification without taking into account positive *as well as* negative potentialities of scalar representations ignores the 'double-edged nature of scale': as both a way to disable alternative forms of valuation by enclosing identities around set borders but *also* as a means of enabling individual agencies. While social scientists must not essentialize rigid social constructs like nationalism, neither should they give in to what Martin Holbraad (2004: 354) calls 'militant anti-representationist analysis', for to do so would be to override the experiences and values of the people who are the very protagonists of social science research. As Anna Tsing argues:

> The best legacies of ethnography allow us to take our objects of study seriously even as we examine them critically. To study ghosts ethnographically means to take issues of haunting seriously. If the analyst merely made fun of beliefs in ghosts, the study would be of little use. (Tsing 2000: 351)

The 'ghosts' in this book are two scalar representations that influence rather than determine how Cuban people make valuations and identifications in everyday life: the global market (which emphasizes individualized consumerism) and the socialist state (which emphasizes the national collective). As we shall see, each has become what Michel Callon and Bruno Latour (1981) call a Leviathan (following the 17th-century philosopher, Thomas Hobbes): ideas and practices that originate at the local scale but which spread across space and time to become powerful and influential human and non-human networks. Though people on the ground also create networks between themselves, their families, their forms of income and their ways of life, Leviathans are more extensive and powerful networks that connect people to wider 'projects of scale-making' (Tsing 2000: 347): 'Two networks may have the same shape although one is almost limited to a point and the other extends all over the country, exactly like the sovereign can be one among the others and the personification of all the others' (Callon and Latour 1981: 280). In this book, the Leviathans are extensive, long-established networks reinforced by 'ghosts' of state and market, whose presence is felt in the creative performances of everyday economic life.

Caveats and limitations

I write from the stance of an anthropologist with budding knowledge of human geography. I hope that this work speaks adequately to both disciplines, particularly by illuminating everyday geographies of food and politics in Cuba. Though I recognize the ethical dilemmas of anthropology by acknowledging the limitations of my own ethnographic lens, here I am more concerned with morality than ethics. This is not just because I am interested in moral economies, but also because of subtle lexical differences between the two words. While 'morality' may refer to personal evaluations of right and wrong, it is also concerned with 'identifications beyond the individual body' (Barnett *et al.* 2005: 5) such as normative appeals to community. Conversely, 'ethics' is often used in reference to more explicit principles for individuals to follow (i.e. the American Anthropological Association's 2012 Statement on Ethics). My choice of one term over the other also fits my subject, for in Cuba morality emerged as a 'political value rather than just a personal ethic' (Kapcia 2008: 92–3). Indeed, as I argue in chapter 2, the durability of Cuban socialism after the fall of the Soviet Union is tied to this long-term moralization of Cuban nationhood.

Given the importance of the nation state for Cuban socialism, the national territory is a starting point for what follows. Such a focus may seem to preclude affiliations that work at other scales, for undoubtedly the Cuban diaspora and international relations of 'solidarity' complicate a one-to-one correspondence between identity and territory. Cuba's 'multiple geographies' (Hernandez-Reguant 2005: 302) are certainly relevant to my analysis, but given the nature of the subject – *everyday* relations between state and market spheres of food provisioning in rural Cuba – my style of analysis must necessarily shift backwards and forwards from the general discourse of territorial unity to the geographical particularities of the Cuban individuals and families under study. Such a technique, based on a dialectical view of representation and performativity, does not preclude analyses at different scales. In fact, the multi-scalar nature of Cuba's food politics is perhaps made *more* apparent as the analytical focus shifts from the national Leviathan to the concrete conditions that shape ordinary lives. Switching from grand narratives to local particularities and back opens up a perspective on the *relations* between scales, which covers more ground than a simple critique of Cuban governmentalities. In line with critical Foucaultian thought, I consider top-down norms as well as the 'lay normativities' (Sayer 2004) through which these are reproduced, if altered or contested, in everyday life.

Despite this focus on the everyday, the book falls short of explaining 'actually existing' possibilities for democratic participation and change in Cuba. The analysis is not framed around the concept of democracy in Cuba (for this, see Roman 2003), for the word 'democracy' itself has a particular history and geography that may not be appropriate for Latin American contexts where the state is understood in different terms (Crabb 2001). If, however, democracy is universally defined as the ability to reason openly in public debate (Sen 2009: 327–70), then, at least ostensibly, Raul Castro's recent call for Cuban citizens to define what 'the Revolution' means to them (2007–2008) or to offer comments and/or critiques about changing policies (2010–2011) (*Información* 2011; *Lineamientos* 2011), is a step in the right direction. I am sceptical of the openness of these public debates, however, for as some of my friends complained of the latter process: 'If you responded with complaints about food … *They* would put you in jail!'

In addition to the above shortcomings, I am also obliged to clarify my use of the word 'culture', particularly in chapters 2 and 5. In revolutionary Cuba the term has been used primarily in reference to a uniform revolutionary Culture (with a capital 'C'), under which artistic, spiritual and folkloric elements are combined into a singular, multi-racial 'cultural patrimony' that defines Cuban identity (*Lineamientos* 2011: 25). My use of the word culture is of a narrower vein, however, as it relates to cultures *of consumption* in pre- and post-revolutionary Cuba. In chapter 2, for instance, I contrast the pre-revolutionary 'high culture' represented by US commodities and the leisure of urban elites with the culture of José Martí and 'Che' Guevara, which I associate with underlying values of revolutionary nationalism like self-sacrifice, hard work and the renunciation of individual desires for collective rewards. In chapter 5, I argue that this counter-hegemonic ideal of consumer culture has become hegemonic in present-day Cuba, providing examples of the way the word is used similarly by ordinary people.

The revolutionary culture of consumption to which I refer in chapters 2 and 5 is also complicated by Marxist views of material progress, according to which 'high' consumerism develops along with the 'productive forces'. Indeed, as I indicate at the end of chapter 2, Ernesto Guevara's idea of the 'new man' of socialism was not only based on a culture of asceticism and hard work, but also on a teleology of future abundance. Moreover, during the 1970s and early 1980s when the Cuban economic model was most aligned with the Soviet model, the government adopted material incentives to stimulate production, creating a 'professional class' that acquired 'high-status household goods' like televisions and home furnishings (Pertierra 2011: 24–5). According to Ariana Hernandez-Reguant (2005: 298–9), 'in the 1980s, research lines [in Cuba] prioritized projects related to the construction of socialism, communist personality, "high" cultural forms of leisure, and consumption patterns'. Nevertheless, from the 1990s to the final period of ethnographic research for this book (in 2011), and likely up to the present, everyday uses of the term 'culture' (chapter 5) are arguably more aligned with the Martian/Guevarian ideal of asceticism and self-sacrifice than with Soviet-style dreams of 'high' cultures of consumption.

Finally, the reader may wonder why I have chosen to order the ethnographic chapters of the body (chapters 3–6) in a consumption (chapters 3 and 4) – distribution/exchange (chapter 5) – production (chapter 6) sequence, rather than the inverse succession as one would expect. The progression – from consumption to production – mirrors the way certain empirical details became available to the ethnographer from the first stages of fieldwork to the last. During the first 6 months or so, issues of consumption were most evident, as my field of inquiry was largely limited to the house in which I lived and the people in neighbouring households whom I befriended. It was only after 6 months living with Cubans that I could start considering rules for distribution and exchange, as it took time for my Cuban 'parents' to reveal comfortably where certain items were provisioned (often illegally), and for the neighbourhood and community to become used to my participation in such practices. The topic of production was the least penetrable to me. Officially and locally, I was recognized as a family member, whose sole 'intention' in visiting Tuta was personal. During fieldwork I was therefore not given access to official statistics on production and imports/exports and I was discouraged from interviewing Tutaño farmers (though I did engage in informal interviews and conversations with them). The data used for chapter 6 (which deals with production) are thus the least ethnographic, although they still required interviews and social time with farmers whom I continued to visit at various periods between 2004 and 2011.

Notes

1   Cited by Wolf (1969: 249).
2   My use of this term aligns with Michel Callon and Bruno Latour (1981).
3   I am indebted to Martin Holbraad for pointing me to work in anthropology that seeks to break down this dichotomy by taking ethnographic realities seriously. Particularly enlightening is his (2004) response to Bruno Latour's non-representational epistemology, which deals with the irony just highlighted.
4   This is comparable to Noel Castree, David Featherstone and Andrew Herod's (2006: 306–7) distinction between 'objective' places, located on a map, and 'subjective' places, which are attached to particular identities (in reference to Agnew 1987).
5   Anthropologists often write in terms of what they call the 'ethnographic present'. A good summary of this analytical device is provided by Mary Douglas and Baron Isherwood (1978: 23), who write that the 'ethnographic present ... synthesizes into one temporal point the events of many periods, the value of the synthesis lying in the strength of the analysis of the perceived present. Whatever is important about the past is assumed to be making itself known and felt here and now. Current ideas about the future likewise draw present judgments down certain paths and block off others. It assumes a

two-way perspective in which the individual treats his past selectively as a source of validating myths and the future as a locus of dreams.'

6   The anthropological project is itself normative, however, for its emphasis on experiential knowledge relies on the assumption that the best way to offer a comparative perspective on what makes life worth living is to live, love and suffer with people for an extended period of time.

References

American Anthropological Association. (2012) *Statement on Ethics: Principles of Professional Development.* http://www.aaanet.org/profdev/ethics/ (last accessed 16 May 2013).

Barnett, Clive, Paul Cloke, Nick Clarke and Alice Malpass. (2005) Consuming ethics: Articulating the subjects and spaces of ethical consumption, *Antipode* 37(1): 23–45.

Callon, Michel and Bruno Latour. (1981) Unscrewing the big leviathan: how actors macro-structure reality and how sociologists help them to do so. In *Advances in Social Theory and Methodology: Towards an Integration of Micro and Macro-Sociology*, edited by K. Knorr-Cetina and A.V. Cicouvel. Boston and London: Routledge, pp. 277–303.

Castree, Noel, David Featherstone and Andrew Herod. (2006) Contrapuntal geographies: the politics of organizing across difference. In *The Handbook of Political Geography*, edited by Kevin Cox, Murray Low and Jennifer Robinson. London: Sage, pp. 305–21.

Crabb, Mary Katherine. (2001) The political economy of *caudillismo*: a critique of recent economic reforms in Cuba. In *Cuban Communism*, edited by Irving Louis Horowitz and Jaime Suchlicki. New Brunswick: Transaction Press, pp. 160–81.

Douglas, Mary and Baron Isherwood. (1978) *The World of Goods: Towards an Anthropology of Consumption.* London: Allen Lane.

Dumont, Louis. (1977) *From Mandeville to Marx: the Genesis and Triumph of Economic Ideology.* Chicago: University of Chicago Press.

Dumont, Louis. (1980 [1966]) *Homo Hierarchicus: the Caste System and its Implications.* Chicago and London: University of Chicago Press.

Dumont, Louis. (1986 [1983]) *Essays on Individualism: Modern Ideology in Anthropological Perspective.* Chicago: University of Chicago Press.

Evans-Pritchard, Sir Edward Evan. (1940) *The Nuer of the Southern Sudan.* Oxford: Oxford University Press.

Fraser, Nancy. (1996) *Social Justice in the Age of Identity Politics: Redistribution, Recognition, and Participation.* The Tanner Lectures on Human Values, Stanford University, 30 April–2 May.

Fraser, Nancy. (2009) *Scales of Justice: Reimagining Political Space in a Globalizing World.* New York: Colombia University Press.

Gellner, David. (2010) Why are the British so peculiar? Reflections on sport, nationalism and hierarchy, *Journal of the Anthropological Society of Oxford* (N.S.) 2(1–2): 31–43.

Gibson-Graham, J.K. (1996) *The End of Capitalism (as We Knew it): a Feminist Critique of Political Economy.* Minneapolis and London: University of Minnesota Press.

Gibson-Graham, J.K. (2006) *A Postcapitalist Politics*. Minneapolis: University of Minnesota Press.

Graeber, David. (2001) *Toward an Anthropological Theory of Value: the False Coin of Our Own Dreams*. New York: Palgrave.

Guthman, Julie. (2004) *Agrarian Dreams: the Paradox of Organic Farming in California*. London and Berkeley: University of California Press.

Hart, Keith. (1986) Heads or tails? Two sides of the coin, *Man* 21(4): 637–58.

Hernandez-Reguant, Ariana. (2005) Cuba's alternative geographies, *Journal of Latin American Anthropology* 10(2): 275–313.

Holbraad, Martin. (2004) Response to Bruno Latour 'Thou shall not freeze frame', *Mana. Estudios de Antropología Social* 10(2): 349–76.

Información. (2011) *Información sobre el Resultado del Debate de los Lineamientos de la Política Económica y Social del Partido y la Revolución*. VI Congreso del Partido Comunista de Cuba. Havana, May.

Jenkins, Timothy. (1994) Fieldwork and the perception of everyday life, *Man* (N.S.) 29(2): 433–55.

Kapcia, Antoni. (2008) *Cuba in Revolution: a History since the Fifties*. London: Reaktion Books.

Lineamientos. (2011) *Lineamientos de la Política Económica y Social del Partido y la Revolución*. VI Congreso del Partido Comunista de Cuba. Havana, 18 Apr.

Massey, Doreen. (2011 [2005]) *For Space*. London: Sage Publications.

OED. (2008) *Oxford English Dictionary*, 3rd edition. Oxford: Oxford University Press.

Pertierra, Anna Cristina. (2011) *Cuba: the Struggle for Consumption*. Coconut Creek: Caribbean Studies Press.

Rankin, Katherine Neilson. (2004) *The Cultural Politics of Markets: Economic Liberalization and Social Change in Nepal*. London: Pluto Press.

Roman, Peter. (2003) *People's Power: Cuba's Experience with Representative Government*. Lanham: Rowman and Littlefield.

Sayer, Andrew. (2004) *Restoring the moral dimension: acknowledging lay normativity*. http://www.lancs.ac.uk/fass/sociology/research/publications/papers/sayer-restoring-moral-dimension.pdf (last accessed 16 May 2013).

Sen, Amartya. (2009) *The Idea of Justice*. London: Penguin Books.

Smith, Neil. (1992) Contours of a spatialized politics: Homeless vehicles and the production of geographical scale, *Social Text* 33: 54–81.

Strathern, Marilyn. (1995) *The Relation. Issues in complexity and scale*, Prickly Pear Pamphlet (no. 6). Cambridge: Prickly Pear Press.

Tsing, Anna. (2000) The global situation, *Cultural Anthropology* 15(3): 327–60.

Wolf, Eric. (1969) *Peasant Wars of the Twentieth Century*. New York: Harper Books.

Young, Iris Marion. (1990) *Justice and the Politics of Difference*. Princeton: Princeton University Press.

Young, Iris Marion. (2011) *Responsibility for Justice*. Oxford: Oxford University Press.

Acknowledgements

This book is dedicated to my two sets of parents – one in Cuba and the other in the United States – for it would not have been possible without the love and dedication of both. The unfaltering generosity and remarkable stoicism of my Cuban 'parents' continue to be a source of strength and inspiration in my daily life. In regards to my own mother and father, I can only say that their long-term commitment to my development as a thinking and caring person enabled me to carry out research for this book against the odds. Along with them, I must thank my partner, Tony, whose thoughtful and meticulous attention to this project – during all its stages – has far exceeded my expectations.

For her thorough and critical guidance throughout the project I am deeply indebted to my PhD supervisor, Laura Rival. Laura saw the potentialities of this research when others did not, and her own scholarly expertise and humanist compassion inspire me to combine the pursuit of career-led research with more humanitarian endeavours. Laura joins other influential women in my life, including the late Sandy Herndon, the late Donna Golding, Laura Nader and my mother, Maryellen Wilson, whose daily encouragement and positivity inspired in me a determination that continues to this day. I am indebted to Stanley Ulijaszek, who continues to encourage me to pursue food and nutrition-related research, and to Neil Coe (human geography editor of the RGS-IBG Series), Peter Jackson, Daniel Miller, Kenneth Olwig, Paul Shaw and my anonymous reviewers for reading through the proposal and/or manuscript and offering helpful advice. I thank my editors at Wiley, particularly Jacqueline Scott (RGS-IBG Series editor), Jennifer Bray (project editor), Jane Andrew (project manager), Eunice Tan (in-house production editor) and Karen Anthony (copyeditor) for their patience and critical guidance. I also express gratitude to the University of the West Indies who provided research funds.

Acknowledgement is made to my Cuban friends and informants (whom I cannot name here) and to those who appear in the photographs: all provided verbal consent for their inclusion in this book. The author and publisher gratefully acknowledge the permission granted to reproduce the following copyright material (in order of appearance): (1) pp. 26–33, 38–40, 43–7 of 'Food as a good versus food as a commodity: contradictions between state and market in Tuta, Cuba', *Journal of the Anthropological Society of Oxford* (N.S.) 1(1): 25–51; (2) pp. 75–80 of '"¡No tenemos viandas!" The political economy of food consumption in Tuta, Cuba', *International Journal of Cuban Studies* 2(1): 73–80 (at the request of the editor, acknowledgement is made to both the editors and to Pluto Press); and (3) pp. 177, 178–84, 186–95 of 'Ideas and ironies of food scarcity and consumption in the moral economy of Tuta, Cuba', *Journal of the Anthropological Society of Oxford* (N.S.) 1(2): 175–98. All of the above were published in 2009 and written solely by this author.

Every effort has been made to trace copyright holders and to obtain their permission for the use of copyright material. The publisher apologizes for any errors or omissions in the above list and would be grateful if notified of any corrections that should be incorporated in future reprints or editions of this book.

Acronyms

AFN	alternative food networks
ALBA	*Alianza Bolivariana para los Pueblos de Nuestra América* (Bolivarian Alliance for the Peoples of Our America)
ANAP	*Asociación Nacional de Agricultores Pequeños* (National Association of Small Farmers)
CCS	*Cooperativo de Crédito y Servicio* (Credit and Service Cooperative)
CDR	*Comité para la Defensa de la Revolución* (Committee for the Defense of the Revolution)
CPA	*Cooperativa de Producción Agropecuaria* (Agricultural Production Cooperative)
CREES	*Centro de Reproducción de Entomófagos y Entomopatógenos* (Centre for the Reproduction of Entomophages and Entomopathogens)
CTC	*Comité de los Trabajadores Cubanos* (Committee of Cuban Workers)
CUC	Cuban Convertible Dollars (the currency that replaced dollars in November 2004)
FAO	Food and Agriculture Organization (United Nations)
FAR	*Fuerzas Armadas Revolucionarias* (Revolutionary Armed Forces)
INRA	*Instituto de la Reforma Agraria* (Institute of Agrarian Reform)
MINAGRI	*Ministerio de la Agricultura* (Ministry of Agriculture)
NAFTA	North American Free Trade Agreement
ONCI	*Objecto Comestible No Identificados* (Non-identified Edible Object)
PCC	*Partido Comunista Cubano* (Cuban Communist Party)
TNC	*TeleNoticia Cubana* (Cuban Televised News)
UBPC	*Unidad Básica de Producción Cooperativa* (Basic Units of Cooperative Production)
UJC	*Unión de Jóvenes Comunistas* (Young Communist Union)

UMC   *Unión de Mujeres Cubanas* (Union of Cuban Women)
UNAH   *Universidad Agraria de la Habana* (Agrarian University of Havana)
WINFA   Windward Island Farmers' Association
ZDAs   *Zonas de Desarrollo Agrario* (Zones of Agricultural Development)

Chapter One
Introduction

When a small household restaurant selling pizza opened in the rural town of Tuta,[1] I was anxious to try it – the daily diet of rice and beans had exhausted its appeal. One afternoon I eschewed household responsibilities like helping my Cuban 'father' peel large numbers of tiny garlic cloves from the state market (Figure 1.1), to try what I hoped to be tourist-quality pizza. All the signs were promising – the ambience pleasant, freshly decorated, cool, the staff attentive, and the food on the menu listed not in national pesos but in hard currency (Cuban Convertible Dollars (CUCs), 1 CUC = 24 pesos = £0.70 or $1.10). I was disappointed. Although bigger and priced at 2 CUCs, the pizza resembled the doughy and tasteless version that sold on the street for 10 pesos, and, like the street pizza, it was made with substandard ingredients, at least according to my own standards of value. Rather than imported luxuries like those available in tourist restaurants in Havana (where hard currency prices are considerably higher), the components of my *napolitano con cebolla* (cheese and tomato pizza with onions) were bought from local sources. Soya oil replaced olive oil, coagulated; artificial cheese replaced mozzarella. There was no wine list.

Now I understand that what was for me a substandard dining experience is for most Tutaños a rare privilege. The long-term relationships forged during the 16 months I spent living in a Tutaño household (from 2005 to 2007 and again in 2011) have given me considerable insight into what it's like to live with scarcity and uneven access (and not just on an intellectual level: I lost 20 lbs!). For most rural (and many urban) Cubans, the restaurant

Everyday Moral Economies: Food, Politics and Scale in Cuba, First Edition. Marisa Wilson.
© 2014 John Wiley & Sons, Ltd. Published 2014 by John Wiley & Sons, Ltd.

Figure 1.1   A comparison of garlic available at the state market (left) and garlic I brought into Cuba from Costa Rica (right). Source: author.

encounter is rare. The infrequency of such occurrences is very likely the reason the state decided to allow a Tutaño family to open the *paladar*[2] in Tuta. Another had been closed down in the area several years before for illegally selling produce earmarked for the tourist industry – in this instance, lobster. *Paladares* are usually geared towards tourists, selling luxury food like lobster at market prices. Most are located in larger cities like Havana. Regardless of the currency in which they operate or their location, all *paladares* are expensive to Cubans as the ingredients for their dishes come from non-subsidized sources. Given the average daily (official) salary in Tuta of 10–20 pesos (about 0.45–0.75 CUCs or $0.50–0.70),[3] neither my Cuban 'parents'[4] nor many other Tutaños could afford to pay 2 CUCs for a pizza. Nor could people I know eat in a tourist restaurant in Havana, where prices are more closely aligned to those of the global market.

Besides those with political capital or remittances, most Cubans do not sit at the same table and share the same food as visitors to Cuba, though the luxuries of tourist life are evident to all. Inequalities of access cause separations between insiders and outsiders, rural and (some) urban dwellers, between people with more political capital and those with less, and between wealthier (and often whiter) residents with remittances and the poor majority who rely on peso salaries. The growing and visible arena outside state distribution networks is the outcome of recent openings of the Cuban economy – particularly the legalization of the US dollar in 1993 (converted to the CUC in November 2008), which now coexists with the domestic peso. While large cities like Havana are already globalized by tourist dollars and 'objective' market determinants, in rural towns like Tuta the vicissitudes of world market prices and global patterns of consumption have only recently (re)entered the social psyche. There, eating at a *paladar* selling food in CUCs is a special event, experienced by an *increasingly visible* minority with access to hard currency. Consumer disparities have not always been so extreme, however. One estimate of the 2001 ratio of highest to lowest incomes is 1 : 70, compared to about 1 : 5 in the 1980s (Fabienke 2001: 65).

In this book, I argue that recent inequalities of access are differentially justified according to one's positioning vis-à-vis internal and external norms and practices or, put more simply, in relation to multiple forms of value. My decision to purchase pizza at a restaurant in Havana – which cost about 10 CUCs, along with another 10 for wine – is justifiable in terms of prices set by a global market combined with the fact that I receive a salary in hard currency from abroad. A functionary of the Cuban communist party might justify taking a larger share of national wealth because he or she works hard for Cuba and sacrifices more (in terms of time, etc.) than others. For a minority of privileged Tutaños, eating at the recently opened *paladar* may be justified because it represents the commercial culture of their envied relatives in Miami.[5] Yet while this so-called 'global' consumer culture may be acceptable in everyday Tutaño life, it cannot be celebrated in open, public spaces. Unlike the tourist, the Tutaño eats behind the curtained windows at the *paladar* (which, one may note, is located far from the main street of town). Here, he or she may *relajar* (relax) in privacy with few fears that official or unofficial onlookers will talk about or sanction such an open act of consumerism.

The above forms of justification are associated with particular scales: the so-called global market, national norms and localized spheres of provisioning. The latter necessarily cut across both national and global economic spaces, and it is the purpose of this book to uncover how such nodes between the local and the global, and between the local and the national, are moralized in everyday life.

National spaces and the logics of food provisioning in Cuba differ markedly from those of the global market, simply because each entails different processes of valuation. My idea of a good pizza evaluated in terms of quality and price – common forms of value associated with liberal forms of globalization – contradicts the ideal of universal distribution within Cuba because, at least officially, such behaviour is associated with individualism and selfishness rather than national solidarity. While the liberal capitalist model is rationalized in terms of the maximization of individual preferences through the ever-encompassing price–profit matrix, the normative 'background' (Barnett *et al.* 2011: 66, in reference to Shove 2003) of economic practice in Cuba centres on a welfare or socialist economy justified in terms of 'non-economic' (Lee 2010: 282) values like the self-sacrifice and hard work it takes to produce collective property. Each political and moral economy is attached to particular values which, over time, become common or lived-in models that guide rather than determine everyday economic life. In defining the terms of 'just' distribution or 'worthy' beneficiaries, each scheme of value (the so-called 'global' economy and the Cuban national economy) also represents a particular culture. Economic materialities do indeed cut across cultural ideas of justice for, as David Graeber has argued: 'What makes cultures different is not simply what they believe the world to be like, but what they feel one can justifiably demand from it' (Graeber 2001: 5).

In Tuta, economic realities are shaped by 'non-economic' values of collective property, hard work and self-sacrifice, which derive from the history of Cuba as a revolutionary nation. As I argue in chapter 2, these values emerged as a significant part of the revolutionary culture in Cuba well before it became a socialist country. Though such values or norms have real-life spatial and material effects within Cuba, they must increasingly coexist with liberal economic forms of value that traverse the protected borders of the nation state. There are, for example, two main material circuits of food in Cuba: those that move within the traditional state-subsidized sphere and those that move through the liberal market sphere. Domestic spaces for food *consumption* are officially separated from tourist spaces, just as spaces for the *exchange/distribution* of domestic goods are morally and practicably detached from those in which imported, more expensive commodities flow. And spaces for the *production* of domestic products are officially separated from those for exports.

Revolutionary values are also implicated in the separation between privatized spaces for privileged consumption and public spaces where acts of provisioning should not contradict 'just' redistribution. Hidden arenas for more individualized consumption are increasingly important, however, as Tutaño families and neighbourhoods 'jump scale' (Smith 1984) from national provisioning spheres to more localized exchange networks. Because privileged Tutaño households – particularly whiter people whose families have left Cuba for Miami – are increasingly connected to outside capital, they may also jump scale in the opposite direction, for a large part of their consumption is met by remittances from, or relationships with, people beyond national borders. The politics of transnational networks increasingly competes with the territorial politics of Cuba as a redistributive nation, and the present task is to reveal how each of these political and moral economic spaces articulate with one another and with localized economic spaces through the 'constant (re)-negotiation of the economic and the non-economic' (Lee 2010: 282) in everyday life. I argue that the long-term values of Cuba as a revolutionary nation influence the way people perceive, spatialize and practise food provisioning in Tuta. As I found during fieldwork, national values, or what I term the moral economy of the Cuban nation (as opposed to moral economies of neoliberal globalization), were *more* influential in performing Tutaño economies than liberal ideas. Tutaño households entered into both state and market provisioning spheres, but it was the former that provided the principal *moral foundation* for economic practices such as the consumption, exchange and cultivation of food.

Market openings in Cuba have led to the re-evaluation and reinforcement of national standards of value, according to which food and the land on which it is grown are forms of collective property. Processes of commodification were not always so morally charged in Cuba, however. Prior to 1959, Cuba resembled other Caribbean societies in its close links to outside commodities and modern consumer values. Memories of this period are still in the minds of some older Cubans I interviewed (see chapter 2), and present-day affiliations with people from the 'outside' (*afuera*) have in some cases allowed for the re-emergence of such forms of value. But since 1959 'food moralities'[6] have been shaped by an inward-looking model of production and consumption, according to which hard work leads to the 'just' redistribution of collective property. Strengthened by historical codes and values, and concretized by their continuous re-enactment, the normative ideal of the national economy is based on the idea of food as an entitlement for all needy Cubans, a form of value not measured in terms of price. This moral economy must coexist with its counterpart, the liberal market, since the one cannot exist without a 'countercyclic' (Polanyi 1944) movement towards the other.

In what follows, I consider political economic values such as price alongside moral economic norms of community, as conceptualized in both state (Cuban) and market (neoliberal) terms. I argue that *all* political economic models, including those at the more liberal end of the spectrum, shape moral, material and spatial ideas of community. Indeed, political economic models may be compared to all other normative schemes associated with identity formation, working as both structure and process, continually in the making.

Political economies: re-connecting 'is' and 'ought'

> For it is not what *is* that makes us irascible and resentful, but the fact that it is not as it *ought* to be. (Georg Wilhelm Friedrich Hegel[7])

As Gunnar Myrdal wrote in his important though oft-forgotten work: *The Political Element in the Development of Economic Theory* (1953): 'Political economy is a grandiose attempt to state in scientific terms what *ought* to be' (Myrdal 1953: 57; my emphasis). Contrary to Myrdal's nuanced account, most understandings of the economy ignore the moral and political foundations of economic models. For Andrew Sayer (2001), this neglect is partly a consequence of a scalar shift in conceptions of the economy from everyday relations based on trust – or the 'Englishman's word'[8] – to anonymous dealings based on self-interest and utility:

As the social relations of economic activity became more functional, anonymous and attenuated and the new economics narrowed its focus to system mechanisms and exchange-value, questions of … value*s* … were gradually expelled. As normative values were expelled from economic science, the scientific or rational content was expelled from descriptions of values, and they increasingly came to be regarded as subjective and emotive. These changes are evident in the vocabulary of political economy over the last three centuries, with a shift from the use of terms like 'virtue', 'vice', 'greed' and 'vanity' to more neutral terms like 'self-interest' (which, of course, can be defined in any way actors choose) and 'utility' that serve to remove economic actions and relationships from critical evaluation. (Sayer 2001: 703; my emphasis)

According to Sayer (2001), positive and normative – or what Hegel called 'is' and 'ought' – have become separated in modern understandings of economy. In most places of the world where neoliberalization prevails, the supra-human 'ought' is a theory of perfect markets and prices, a kind of 'virtualism' (Carrier and Miller 1998) that works as both a model *of* and a model *for* (Geertz 1973: 87–125) actual economic relations. In derivatives markets, for example, the 'virtual' value of an asset becomes actual profit when a number of buyers *trust* predictions of its future value.

This kind of performativity, according to which the 'ought' becomes 'is' through its re-enactment, is similar for the Cuban socialist model in which transcendental values, such as the idea of society comprised of 'new men' of socialism (see chapter 2), are also dialectically related to real-life outcomes. Like the theory of perfect markets, which continues as an 'ought' despite recurring global financial crises, the Cuban Communist Party (*Partido Comunista Cubano*, PCC) continues to re-develop old values in new contexts. Thus, while the legalization of hard currency marked a break from the socialist economic model, the latter has remained dominant in official circles. Officially, food (like healthcare and education) should be treated as an 'inalienable'[9] or non-commodifiable public good, produced by and distributed to citizens according to the communist tenet: 'From each according to his capacity, to each according to his needs' (which has been significantly altered in recent years, see chapter 4). Accordingly, pesos embellished with pictures of revolutionary heroes are treated primarily as 'tokens'[10] earned through hard work and exchanged for provisions which the state defines as necessary such as rice, beans and eggs. Yet this model stands in moral contrast to the growing hard-currency economy in Cuba, in which food of a higher quality than that provided by the state is treated as a *commodity* only accessible to a few and acquired through more individualized market transactions.

I argue in chapter 4 that the ideal moral economy in Cuba is now being reworked by President Raul Castro and his government. Not unlike recent calls for individualized 'responsibilization' in liberal capitalist countries, new economic plans in Cuba are aimed at lessening the burden of state redistribution by emphasizing workers' responsibility to produce collective

Figure 1.2   The state-subsidized market in Cuba (compare to the private farmers' market shown in Figures 1.3, 3.2 and 5.5). Source: author.

property (González Maicas 2012). President (Raul) Castro advocates a slow but progressive rationalization of the state workforce and a progressive change from universal rationing to a kind of means testing for benefits, similar to the neoliberalization of some countries in eastern Europe (see Stenning *et al.* 2010: 3–5, 175–80). Unlike eastern Europe of the 1990s, however, dominant values of the Cuban revolution – nation, self-sacrifice, hard work and collective property – are, at least at an official level, largely left unchanged.

In post-1990s Tuta, scarcities and daily hardships have blurred the boundaries between morality and opportunity, legality and necessity. Under these conditions, the moral and jural border between collective morality in the state sphere and entrepreneurial tactics for acquiring scarce commodities in the market sphere, has become more fluid. Scarcities combined with market openings have given rise to ambiguities at the local level between 'is' and 'ought'. Such moral uncertainties allow Tutaños to find a space for inventing new rules to legitimate legally dubious activity. I outline such local rules explicitly in chapter 5, borrowing from an idea originally devised by anthropologist Janet Roitman (2005) in her account of the informal economy in the Chad Basin. Roitman's reference to 'licit-illegal' (Roitman 2005: 204) or what I call illegally licit behaviour is appropriate for my own analysis of local tactics for acquiring scarce commodities, though for different reasons. While Roitman's account of the way people legitimize otherwise illegal or immoral activities in Central Africa refers to the acts of deviants, in Tuta, illegally licit behaviour is necessary for most, if not all, Tutaños. Because of the frequent need to break legal rules, Tutaños must find *moral* ways to resolve contradictions between state and market spheres of provisioning. Indeed, as I emphasize, Tutaños justify otherwise unacceptable economic practice by drawing from underlying values embedded in the history of Cuba as a revolutionary society.

Shifting scales of responsibility

Like the national moral economy in Cuba, which rests on ideas about the obligations and legitimate undertakings of state and citizens, everyday moral economies in Tuta are underpinned by obligations between a person and his or her extended family and/or neighbourhood. Indeed, while the *official* boundaries of community in Cuba are national borders, in Tuta 'community' is often limited to one's *familiares* (extended family members). Despite cultural continuities, scales of affiliation and responsibility have changed significantly – from responsibilities and obligations between state and workers, legitimized in material terms by peso salaries exchanged for necessities, to those between families and workers, whose low salary in pesos must be augmented through legal or illegal market activity or gifts from people *afuera*. The state's role as provider has changed from a somewhat trusted guardian to a distant other, epitomized by the designation 'They'. As one Tutaño put it, 'you must take what you can get from *Them*'.

A key moral dilemma in present-day Cuba is this jump in scale of provisioning from the nation to the locality, and it is a central issue explored in this book. Recent changes in the materialities and spatialities of food provisioning, from the state sphere to more localized (or transnational) exchanges with one's (extended) family have caused a shift in responsibility from the nation to the family and/or neighbourhood. At the same time, more and more people are attached to family outside Cuba. In 2009, US President Barack Obama passed a law that increased the limit for remittances to $3000 for every visit (from $300 per quarter during the Bush era). With unlimited travel permissions (now, including travel *from* Cuba),[11] Cubans and Cuban-Americans may visit friends and acquaintances as well as immediate family members. There are no longer any weight restrictions on the amount of luggage that can be brought into Cuba. This has facilitated an influx of US commodities, and for some, a taste of modernity.

Network communities are partially replacing territorial communities in Cuba and elsewhere (see Miller 2011). Still, supranational connections have always influenced Cuban identities. A mythical Cuban hero like José Martí (see chapter 2) lived in the United States for over half his life, and since 1959 many Cubans have travelled with the military or, more recently, volunteered for medical 'missions'. Others have simply left Cuba for a 'better' life in the United States or Europe. What makes present transnational relations between Cubans and their relatives living *afuera* different from these other experiences is that they have created *increasingly visible* economic disparities between people linked to outside markets and people largely dependent on the state redistributive system. Indeed, as a consequence of the opening of the Cuban economy and ensuing inequalities, the historical dichotomy between revolutionary sympathizers and so-called individualistic 'traitors' (Whitehead 2007: 14) is perhaps more internalized now than at any time since the beginning of Fidel Castro's revolution.

As the ratio between the highest and the lowest incomes in Cuba has shifted, so the state is no longer the primary provider for its citizens. Like other places, in Cuba globalization has partially ruptured what Nancy Fraser (2005, 2009) calls the 'Westphalian' sovereignty of the nation state.[12] Similar to processes of globalization elsewhere, public goods in Cuba are increasingly eclipsed by private commodities; redistributive justice and economic policies are no longer entirely under the remit of national economic controls. Wages in Cuba are so devalued in terms of the global market that they have largely lost their status as tokens that may be exchanged for necessary household provisions. The ideal political economy in Cuba, which hinges on concepts such as public goods and redistribution, is now at odds with alternative ways to provision for the household, some of which are connected to the global political economy. In these new circumstances, economic responsibilities for the local family (often met by extranational networks) must coexist with or even supersede moral or political responsibilities to the national 'family'.

In Cuban terms, the dialectical relationship between national and local (/extranational) scales of responsibility is perhaps best reflected in the concept *'lucha'* (struggle or fight, here I mostly use the latter translation), a common word and cultural construction that has already been identified in other studies of Cuba.[13] Unlike most scholarly accounts of the concept, however, I argue that *lucha* has two separate often *opposing* connotations, which correspond nicely to the distinction between national values and local economic realities. In the national moral economy, *Lucha* (with a capital 'L') recalls the revolutionary 'Fight' against imperialist exploitation, mercenary selfishness and unjust ownership of national resources that have sparked revolutionary ideas in Cuban society at least since the late 19th century. As chapter 2 illustrates, it is necessary to consider such representations of the nation state since they become everyday tools for 'social and political projects that underwrite discourse and performance... stress on one repertoire over another can affect the outcome of power struggles, opening up opportunities to one set of claimants, foreclosing them to another' (Wolf 1999: 8).

In Cuba, the word *'Lucha'* is a cultural conception tied to the historical 'Fight' of the Cuban nation against those who use (or have used) national wealth for mercenary purposes. In this cultural scheme, the national *Lucha* against such 'traitors' will lead to a just society, where food and other social

property is redistributed from the centre to workers who receive their just
due according to their level of dedication to the revolutionary cause.Yet this
version of *Lucha* has historically overshadowed other ideas about provisioning which emerge in local contexts, reflected, for example, in the positive
memories some Tutaños have of consuming commodities from 'the North'
(i.e. the United States). As Caroline Humphrey (1989) argued for the Janus-
faced use of Soviet terminology by a Mongolian people in south-eastern
Siberia, representations that are 'uni-accentual' within structures of power
are made 'multi-accentual' by social actors within local contexts (Humphrey
1989: 145–6). In Cuba as in other contexts of nationalism, '[t]he more fixed
the semiotic forms, the greater is the play of ambiguity and the more surprising are the possibilities for violating the code itself' (Herzfeld 2005: 20). At
more localized scales, where the boundaries of community may span the
area of one's neighbourhood as well as non-territorial relations with distant
relatives, the word *lucha* has come to mean the daily 'struggle' or 'fight' to
survive long-term scarcities that have characterized Cuban life since the
early revolution, but especially since the 1990s when Cuba could no longer
rely on barter terms of trade with the Soviet Union (see chapter 3). '[T]o
utter "I am struggling" (*estoy luchando*) is to communicate in one verb the
years-long efforts and defeats in small acts of life' (Pertierra 2011: 81).

   Moral and political differences between the national *Lucha* and the local
'*lucha de provisiones*' (fight for provisions) have not been overlooked by the
Castro brothers and other top officials in the Cuban government.The latter
criticize the 'double morality' (*doble moral*) of some Cubans: a so-called
hypocritical switching that occurs when a person openly commends the
ideals and values of the Revolution on one occasion, while engaging in economic acts to further his own 'interests' at another point in time (Suárez
Salazar 2000: 245).Yet, as noted above, contradictions characterized by the
doble moral concept, or the need to balance the requirements of national and
everyday moral economies (*la Lucha* and *la lucha*), are an underlying factor
of ordinary economic life in Cuba. While many Tutaños I met openly
defended national goals to create and protect social property through work
and dedication, most also felt the need to engage in 'the fight for provisions', often by breaking official rules by engaging in market transactions.
Anthropologist Anna Cristina Pertierra has usefully referred to this as the
'entrepreneurial-humility dialectic' of Cuban life (Pertierra 2007: 146).

   Even officials – the primary representatives of the state (Abrams 1988) –
break formal rules in order to obtain commodities that are otherwise only
available as public goods distributed through state channels. Indeed, just as
there are spaces in-between state and market values, so there are interstices
that complicate strictly Foucaultian views of Cuban governmentality. Such
spaces in-between state and civil society are plainly illustrated by the words
of one Tutaño who lamented how state managers' morals had changed since
the period of extreme scarcities in the early 1990s:

A: There has been a change in values since the late 1980s. Now there is so much crime. People steal from the state or wherever and it doesn't seem to matter anymore. Before, no one stole. When they did, there was a moral outcry. Now people want good state jobs because they want to get things from the state. There are not as many restrictions as before.

M: Why is that?

A: Because the *jefes* [state bosses or managers] don't report anything. They take things also.

Crossovers in anthropology and geography II

This book responds to the need voiced in geography for empirical evidence to unravel the political potentialities of everyday spaces in-between top-down and bottom-up forces. In addition to anthropological theory and method, geographer Roger Lee's (2006) concept of 'ordinary economies' or more appropriately, 'ordinary economic geographies', will be essential for this project. As Lee argues (2006: 413), the economy 'is an integral part of everyday life, full of contradictions, ethical dilemmas and multiple values that inform the quotidian business of making a living. In short, it is ordinary'.

The added spatial dimension when 'ordinary economies' become 'ordinary economic *geographies*' adds to understandings of everyday value formation in economic anthropology since it points not only to multiple value systems but also to various spatial trajectories flowing within and without the borders of economic 'units'. Though the idea of 'coeval' relations of value is a cornerstone of economic anthropology (e.g. Gregory 1997), and though the spatialities of contradictory 'spheres' of value are central, if often implicit, to its theoretical history (e.g. Bohannan and Dalton 1962), the 'who' of such relations have usually been defined in terms of bounded economies/communities, such as the *Tiv Economy* (Bohannan and Bohannan 1968).[14] In Nancy Fraser's (2009: 39) terms, the 'frame' of economic activity was largely determined by the social scientist, who maintained it by a strict dichotomy between traditional communities and outside economic forces.

As a result of the 1960s–1980s theoretical debate in anthropology between substantivism and formalism,[15] more recent work in economic anthropology has overturned the dichotomy between non-market and market values, uncovering interrelations between different forms of value in everyday life (in traditional as well as modern societies) and revealing the importance of the market for non-market relations (e.g. of care) and vice versa.[16] In the light of this literature, usually transferred to geography via Daniel Miller (who has not escaped criticism in some anthropological circles),[17] Lee (2006: 413) uses the concept of 'ordinary economic geographies' to disclose the moral and socio-political contradictions of everyday economic life. While adding to the range of sources in anthropology that interrogate multiple forms of value in everyday life, Lee's concept of ordinary economic geographies also uncovers the multiplicity of *spatial* trajectories that link (or do not link) inside and outside forces. Treating economic geographies as ordinary and full of contradictions not only allows the theorist to escape a unifying logic that sublimates one form of (political and moral) economy to another, but also frees economic activity and economic actors into a so-called 'heterospace' (Gibson-Graham 1996: 5) of diverse internal and external relations.

Much of this book provides evidence for such a perspective. Indeed, in a political sense, I view the Cuban economy as an open space, full of various possibilities. In an empirical sense, however, I do not ignore the performative role of economic models – that is, how normative designations of 'the econ-omy' become 'lived-in models of the world' (James 2003: 57) that really do *matter* to people.[18] In line with recent work in geography (e.g. Barnett *et al.* 2011), I argue that dominant economic norms of distributive justice in Cuba affect everyday commitments and guide economic decisions. Yet individuals and groups necessarily shift between dominant conceptions of value and their associated spatialities and others, including those underpinning the global market economy. While I am careful not to frame my argument in terms of any one particular unit (i.e. the national economy, the local community), in relating theory to practice, scale still matters. For the values circumscribing particular scales have historical and spatial continuity as they are adopted, altered or contested in everyday life. Thus it is often the case that

> one scale – the world or global economy scale, or the national, state-building scale – 'really matters' when counter-poised to local struggles. ... This would be too closed, too close to a Foucaultian web of domination which fatally under-mines agency if we were to forget that there is no set, unchallengeable hierarchy and that space *is* constructed from a relational mix of scales (Taylor 1999) in changing and challengeable scalar political opportunity structure (Tarrow 1998). ... But actions at some scales have more influence than others in some conditions, and some scales are less open to contestation than others. (North 2005: 225)

In Cuba, political economic (and moral) influences from without are 'contained' (North 2005: 224) at the national scale, for the state controls most places where commodities flow and redirects profits from outside market relations (from the tourist sector, for example) to national coffers (which are then, supposedly, redistributed justly). In Michel Callon and Bruno Latour's (1981) terms, the socialist government in Cuba, with all its human and non-human resources, has become a Leviathan that sets domes-tic food apart from the outside world of commodities. The discursive and material practice of 'enclaving' (Appadurai 1986: 22–6) in Cuba – or separating things out from the realm of commodities – is justified by the transcendental notion of a future transition to a Cuban version of socialism. Professor of Cuban history, Antoni Kapcia (2000), associates this teleology with what he calls an 'ideology of dissent' or '*cubanía rebelde*' (rebel Cubanness) that historically emerged in the face of outside imperialist interests, a trajectory I briefly trace in chapter 2.

The counter-discourse to *cubanía rebelde* – the neoliberal transition of Cuba – also 'imposes its own space and time' (Callon and Latour 1981: 286) by asserting a geographically and historically uniform path for economic liberalization. Both Leviathans are underpinned by the claim that economic activity within the border (the nation or the globe) will transform in uniform ways. Thus according to a neoliberal view, Cuba's positioning within the North American 'sphere of influence' will inevitably lead to its integration into 'the' global consumer culture, unifying present incommensurables into the universal price mechanism. In an opposing scheme, Cuba (and perhaps others in the region) will inevitably follow the path of a socialist future, under an entirely different framework for distribution and justice, but unified by national borders (at least for now).[19] Each moral and political economic framework is attached to a particular scale, and the scalar politics of each 'claim[s] a privileged perspective' (Gregory 1997: 18) that defines the parameters of 'the economic'. Destabilizing state and market discourses opens up political possibilities simply because dominant economic logics such as the 'right' price for a commodity or the necessary '*perfeccionamiento*' (advancement) of the socialist economy (Castro Ruz 2010) become historically and geographically contingent, rather than natural or immutable. Relations between a national politics of scale, associated with the view that food is an entitlement, and a global politics of scale tied to the idea of food as a commodity, are negotiated by ordinary people on the ground according to social positionings vis-à-vis inside and outside forces. Indeed, ordinary economic geographies in Tuta involve a multiplicity of capitalism*s* (Gibson-Graham 1996) and socialism*s*, with political potentialities that are not captured by stark binaries between state and market or 'alternative' and 'mainstream' systems of value.

Capitalist as well as socialist, economic as well as non-economic logics are reproduced within the 'whole' of economic life in Tuta. In this book, I relate empirical activity experienced, discussed and observed in Tuta to the various 'wholes' working within and between people, in which both sides of polemic debates play a part.

[D]istinctions [between individualism and collectivism] are numerous, fluid, flexible, running independently of each other, overlapping or intersecting; they are ... variably stressed according to the situation at hand, now coming to the fore and now receding. On the other side, we [moderns] think mostly in black and white, extending over a wide range clear either/or disjunctions and using a small number of rigid, thick boundaries defining solid entities. (Dumont 1986 [1983]: 253)

Figure 1.3   A scene from the *Mercado Agropecuario Campesino* (Free Farmers' Market) in Tuta.
Source: author.

While many people still work for the collectivist state, most must also
engage in individual pursuits like buying candy from CUC stores and selling it for slightly more in pesos. In the everyday *lucha* to find a balance
between national and domestic needs and demands, it is, as one man put it,
'always important to have two jobs, one for the state and another on the
side'. The job 'on the side' is usually legal or illegal petty trade in the limited
market sphere, such as the sale of home-ground coffee, the coffee seeds
having been 'smuggled'[20] from another municipality or province.
Nonetheless, the ultimate source of such petty commodities often derives
from the state sphere: a state worker or *jefe* 'grabs' (or 'steals'; this distinction is explained in chapter 5) goods from a state-led distributive institution
and then sells them to an intermediary.

Even if one operates strictly within the law and purchases from official
outlets (e.g. the Free Farmers' Market, Figure 1.3), it is still difficult for
Cuban buyers to know whether vendors have met all the necessary legal
requirements: 'When Elisabeth buys okra from vendors, who is to say
whether the vendor has paid her monthly fee?' (Pertierra 2011: 137). Like
anthropologists Norman Long and Paul Richardson (1978) argued for
Peru, and as geographer Adrian Smith (2002) argued for Slovakia more
recently, non-market practices in Cuba often support and reproduce capitalist relations. Nevertheless, I argue in the ethnographic chapters (3–6)
that the various capitalism*s* in Tuta are *moralized* in terms of long-term
values of Cuban nationalism (and, now, socialism) such as selflessness,
struggle and familial solidarity.

Anthropologists have long argued that everyday shifts between different 'spheres of exchange' are always morally charged. While the spatial contours of such shifts have often been left out of anthropological accounts, geographers have only recently begun to use ethnographic material to consider shifting and heterogeneous economic logics in everyday life (e.g. Stenning *et al.* 2010). Here I draw from both disciplines to reveal how people in rural Cuba rationalize the practicalities of living within and between contradictory, though coeval, economic and discursive spaces. In accordance with recent work in geography (Leyshon *et al.* 2003; Williams 2005; Fuller *et al.* 2010), I counteract stark contrasts like that between neoliberalism and Cuban socialism, or between a 'mainstream' economy and an 'alternative' communitarian ideal, which obscure actually existing multiplicities in social, spatial and economic life. In accordance with anthropology, however, I recognize the empirical significance of such binaries (see also Samers and Pollard 2010: 49). Indeed, as we shall see, binary oppositions between 'inside' and 'outside' in Cuba are used by officials and ordinary people alike to define outsiders *within* who, in the dualistic ideology of Cuban nationalism-socialism, conform too closely to the individualism of capitalist markets (see chapter 5). On the ground, 'outside' macroeconomic geographies often collide with 'inside' 'spaces of political engagement' (Jonas 2010: 21). This does not mean, however, that what we have come to see as 'the economic' is an apolitical phenomenon, a force that comes from nowhere.

The economy–culture relation

Economic concepts like capital, supply and demand and so on, are systematic elements that affect every place in very real material ways (Sayer 2001). But that this is so in our present era has more to do with the foundations of these principles in particular politico-cultural discourses, which have become empirically valid through their on-going performativity, than because of any economic determinism premised on their spatial and temporal inevitability. Like other forms of representation, there are two aspects of 'the economic': an empirical referent and its discursive association with the dominant cultural politics of late capitalist society, particularly the dominant vision of neoliberalization. For those striving for alternatives, it is unfortunate that 'part of this [discursive] articulation [is to] quietly … englobe the former function' (Dresch 1976: 57).

Perhaps less unfortunate is the realization that 'global' neoliberalization has a particular genealogy (see Peck 2010) that prevents its uniform incorporation into *all* cultural worlds, including the dominant (and often oppressive) agenda in Cuba. Relations between this (macro)economy and place-based politics and cultures are therefore always historically and geographically contingent:

> We do not have to flip from the dogma of the economy as determinant in the last instance to the dogma of culture going all the way down. There is no sense in making general pronouncements on their relative importance, since this is always an empirical question that will depend on the particular case in which it arises. (Sayer 2001: 705)

I argue in chapter 2 that the long-term opposition between socialist and/ or nationalist values in Cuba and (neo)liberal values in the United States has, if anything, strengthened cultural 'spaces of dependence' (Cox 1998) within Cuban territory rather than annihilated them. As indicated above, appeals to social unity in Cuba often rest upon the identification of outsiders residing within the borders of the nation state, a situation that recalls anthropologist Anthony Cohen's argument that the 'social identity of a group may ... be contested *within* the group itself, on grounds related to the *cross*-boundary interaction' (2000: 1; emphasis in original). Thus Cubans who are seen as having too much money or too many luxury goods *without* making appropriate sacrifices for the benefit of the community face both official *and* unofficial demands to realign the economic balance between themselves and 'all' Cubans. Understanding such cultural norms is essential for scholars interested in the culture–economy relation, for as anthropologist Timothy Jenkins argues, '[o]fficial claims as to the nature of the world are part of the material to hand, with which, from an actor's perspective, to assert worth and to act in the present' (Jenkins 1994: 451).

In Tuta at least, most people do not participate in a community of 'common ends' (Young 1990: 229) like that upheld by the 'alternative-oppositional' (Fuller and Jonas 2003: 66) stance of the Cuban government. Many identify with people in their neighbourhoods in terms of collective imaginings of a better future and a shared sense of caring for others, often using cultural forms that derive from nationalism and/or socialism. The lack of political dogmatism at the local level – apart from official circles – is perhaps illustrative of the kind of 'heterogeneous unities' Iris Marion Young (1990: 236) referred to as part of her dismissal of binary political economic thought. Like Young, my aim is to 'explode the binar[ies]' (Gibson-Graham 1996: xxi) between culture and economy, state and market, socialism and capitalism.

Rarely does a visitor to Cuba have the kind of access that enables an understanding of the bigger picture that contains all these binaries and everything in-between. During all my visits to Cuba, excepting my time in Tuta, I have either stayed in tourist accommodation (such as *casa particulares*) let by wealthier Cubans who usually receive remittances, or else I have been compelled to lodge in official residences for foreign academics where the level of economic activity one is allowed to witness is highly restricted. My only gateway into rural society has been through the creation of a fictitious kinship relationship with my Cuban 'parents'. Thankfully, this relationship was recognized both officially (through family visas) and unofficially (in the neighbourhood) as a valid form of belonging and habitation, for without such recognition I would have experienced very little 'ordinary' economic life in Cuba.

Positioning the ethnographer I: habitual and representational knowledge

Apart from interviews and the collection of life histories (as well as historical and discourse analysis, see below), the principal methodology used in this book is ethnography. While the word 'ethnography' is now used in loose terms, referring to a period of research that usually does not last for more than a few months, any period of time spent living in a different cultural world enriches the field of knowledge, allowing the researcher to take a reflexive stance and perhaps leave off some of her own cultural assumptions in the analysis that follows. Positioning the researcher in the cultural field thus

> offers a dynamic, agentive model of identity construction where a person creates a possible identity for themselves in a particular context through their active positioning in relation to, or perhaps in opposition to, elements in their discursive cultural context. (Linehan and McCarthy 2001, cited by Barnett *et al.* 2011: 121)

While not 'emplaced' (Mansvelt 2005: 85) in the same way as the people under study, the ethnographer is 'committed in the body' (Jenkins 1994: 451) and so acquires the kind of knowledge – both habitual *and* representational – that is particularly useful for studies that highlight food scarcity and related hardships. I will take these two forms of knowledge in turn.

Habitual knowledge may be acquired through participation in communal activities such as preparing and eating food, watching television and contributing to conversations about particular programmes, cleaning the house and engaging in other forms of 'work' (reproductive work is recognized as 'real' work in Cuba), dancing and drinking rum – all of which were essential for this study. But understanding habitual behaviour goes deeper than daily tasks and shared social events; it also requires openness to making mistakes and learning from them. Such resolutions become apparent only after the ethnographer internalizes cultural 'rules' for behaviour that are often implicit. This is usually only possible after significant periods of time spent living in a household, loosely defined as any combination of people living under one roof. Thus as I learned, it was offensive to bring home food or drink for one's own consumption without offering some to family members, it was considered 'dangerous' to spend time with certain people or to give gifts of things like beer in certain places (see chapter 5), it was considered anti-social not to sit through the midday hour of soap operas, even if I thought it was a waste of time.

Such slight forms of information are often crucial, for when the ethnographer learns to act appropriately he or she may establish trusting relationships and perhaps gain access to social spaces (like informal exchanges) that illuminate more profound insights into daily life. For this reason, the ethnographer spends much time taking notes, perhaps summarizing them, periodically, into summary essays, as I did about once every few weeks whilst living in Tuta. The practice of note-taking, along with others like developing memory skills (I would recall 5–7 main topics of hour-long conversations, which later helped me remember details in between),[21] creates a kaleidoscope of knowledge about the field site that may, as the metaphor implies, change over time. Though this vision is never (and never can be) complete, the pieces thus arranged can give the ethnographer increasing insight about what really 'matters' to the people with whom one lives.

While some may critique the value anthropologists place on habitual knowledge as 'deliberately render[ing] their research subjects mute' (Barnett *et al.* 2011: 78), most anthropologists take verbal 'forms of justification' (Boltanski and Thévenot 2006) very seriously indeed. By recording common forms of justification repeated by various people over an extended period of time, the ethnographic researcher may establish plausible if always *provisional* and *testable* (Jenkins 1994: 434) conclusions about the ways those people conceptualize and practice the 'good life':

> As soon as the researcher can no longer base the validity of her affirmations on her stance of radical exteriority, the definitiveness of the description comes into question. In such cases the researcher is obliged, in her description, to adhere as closely as possible to the procedure the actors themselves use in establishing proof in a given situation; this approach entails paying careful attention to the diversity of forms of justification. (Boltanski and Thévenot 2006: 12)

Representational knowledge, such as justifying economic practice with words like '*lucha*', may be as important as habitual knowledge. In places like Cuba, such political discourses are often incorporated – whether in earnest or in play – into everyday social exchanges. My ethnographic notes included such non-verbal mishaps as those listed above, but I also took note of the frequency of politically charged words or ideas and of the contexts in which they were used. The ethnographic context demanded that I compare everyday usages of expressions like '*lucha*' or someone with or without 'culture' to their official and historical counterparts. I did so through the collection of histories, newspaper articles, policies and speeches, modes of data collection that will be evidenced in what follows. But it was ethnography that allowed me to understand the relations (and contradictions) between discourse and practice, avoiding a kind of 'militant anti-representationalis[m]' (Holbraad 2004: 354) that ignores the ways political norms enter into the 'world of concern' (Sayer 2011).

Positioning the ethnographer II: food and the 'politics of negotiation'

Given the embodied nature of experience and the workings out (or not) of difference and privilege through experiences like consuming food, the ethnographer becomes an important player in the very 'politics of negotiation' (Massey 2004, 2011 [2005]) he or she may be studying. Geographers of food have recently claimed that food is politically important because differential access to food is experienced as an embodied form of exclusion. Food is a special kind of object in both its commoditized and non-commoditized forms. As a commodity – something with an exchange value or price – it stands in stark contrast to other commodities like tablet computers. It is not only physiologically essential, but also a powerful symbol of social conditions such as 'luxury and lack' (Mansvelt 2005: 95), dependence and autonomy. As a non-alienable good, food may be used to define and (re)-create community. As such, food is an important lens through which sociopolitical relations may be studied:

> Food stands in a different category from the ordinary commodities of economic exchange. It is an insistent human want, occurring regularly at short intervals, and shared by the whole community alike. ... [I]t is the mechanism by which food-getting habits are formed in the structure of each different culture that we have to analyze. (Richards 1932, cited in Wilson 2012: 278)

Consuming food, in particular, opens up connections and/or disconnections between the researcher's own field of experience and that of the people with whom one interacts. As Ian Cook et al. (2010) argue, 'foods link up with ideas, memories, sounds, visions, beliefs, past experiences, moods, worries and so on, all of which combine to become material – to become bodily, physical sensation' (Cook et al. 2010: 113–14). Understanding such differences and multiplicities – in habitual *as well as* representational terms – allows the fieldworker to distinguish his or her own experiences and knowledges from those of the people encountered during fieldwork, and to reach normative or political conclusions.

A face-to-face politics is engendered by a meeting up of different life trajectories, positioned differentially in relation to 'power geometries' (Massey 1991) that separate privileged from underprivileged spaces of consumption, for example. Though my own spatial and moral positioning in rural Cuba was far from that of the people with whom I lived, like them I had to shift from one trajectory to another, in terms of my *own* relation to inside and outside power geometries. And like my 'informants' (mostly friends), I was able to step into and out of different economic and moral roles according to the specific endo- and exo-relations brought about by the encounter. For instance, I shifted from being an ethnographer, tolerantly experiencing rural life with all its hardships, to being a tourist entering privileged spaces of consumption just like any other rich tourist in a poor country would. Indeed, this is how I justified it to myself! My own experiences as a tourist, attached to memories of other places and times, provided the economic and moral 'background' with which such a justification became reasonable. As a member of 'the' global consumer culture, I simply needed to satisfy culturally established needs – like decent pizza – through periodic trips to Havana.

Entering into tourist spaces counteracted the stoic stance of the ethnographer who is supposed to 'go native'. But this naive methodology, which assumes that people can simply re-localize themselves and their values to fit local understandings, later describing these to others through representation, reflects a flawed notion of an 'authentic' fieldwork experience. It is also related to a cultural belief in an objective 'truth' rather than multiple versions of shared experiences.

Geographers like Doreen Massey extol the 'radical contemporaneity' (Massey 2011 [2005]: 99) of multiple trajectories that link up (or do not link up) persons with different histories and ideas about the present and future, and with different spatial relations to phenomena like economic globalization. Anthropologists should surely appreciate her normative appeal to recognize the political nature of space as well as time, for ethnographic knowledge consists simply of 'potentially exclusive versions of the truth that together constitute the event' (Jenkins 1994: 443). Discrepancies between the ethnographers' own experiences and values and that of his or her informants are actually helpful in establishing Massey's politics of 'coexisting heterogeneity' (Massey 2011 [2005]: 9), though of course we cannot take this so far as to ignore power relations that work at scales beyond the individual.

The provisioning perspective

Cuba's changing economy, agroecology and society offer social scientists a unique opportunity to apply theories of responsibility, justice and value to everyday life. Starting from my own experiences in Cuba, which first came to light when I encountered separations between tourist and Cuban spheres of consumption, in this introductory chapter I have laid the foundations for the central methodological and theoretical premises of the book. Though it necessarily follows a linear scheme (from consumption to production), my analysis is non-linear in its treatment of provisioning processes. I treat each provisioning process – production, exchange/distribution and consumption – as connected not only through shared cultural understandings of appropriate economic practice (Fine and Leopold 1993; Fine 2002; Narotzky 2005) but also through particular material networks maintained through Cuba's unique system of governance and power. I am particularly concerned with processes of commoditization, and to differences between different kinds of commodities (Crang *et al.* 2003: 448), identifying food as an object through which both commodified and non-commodified provisioning processes are morally embedded.

Similar to other anthropologically informed studies in geography, I treat decommoditization as a counterpart to commoditization, and vice versa, a value relation that must be worked out in practice. Unlike some in the geographical tradition, however, I contest the idea that commoditization 'breaks down' the moral economy (Watts 1999: 310; for a critique, see Williams 2005), or that the political economy of commodities must always undermine forms of exchange that are more 'embedded' in moral economies. Indeed, I consider state and market theories of value, and their concomitant socialities, spatialities and materialties, as moral economies in their own right. If 'entanglements' and 'disentanglements' between persons and things are equally human (Callon 2005: 6), then the analytical use of concepts such as commodities should not precede considerations of the social relations through which they become represented, understood and treated (or not treated) as such. Like Clive Barnett, I start not from the commodity concept per se, 'the "software" of using and exchanging goods', but the 'socio-material "hardware" that supports such activity', or the 'systems of provisioning and background infrastructures that enable all these affective exchanges to go on' (Barnett *et al.* 2011: 66). Unlike a more commodity-centred approach, then, I emphasize processes of value formation that underpin the systems of provision approach (Fine and Leopold 1993; Fine 2002; Narotzky 2005), attempting to uncover the 'complex ways that power and interest can shape a provisioning chain' (Narotzky 2005: 83–5).

Outline of the chapters

As emphasized, the national moral economy that structures (rather than determines) systems of provisioning in Cuba stands in direct contrast to liberal economic thought, which treats food as a commodity accessible to anyone with enough hard cash, regardless of their moral standing. In chapter 2, I explain the contrast between the normative idea of food as an entitlement upheld by the Cuban nation state and the unlimited desires of consumer capitalism, and uncover how each creates normative links between consumer and community, citizen and nation state (or globe).

I relate the Cuban national moral economy to the historical and geographical context in which it developed, positioning Cuba as an alternative economic space vis-à-vis its 'mainstream': the United States. In this way, I assign a 'home' (Massey 2004: 427) to an otherwise contingent project of national unity in pre-socialist and socialist Cuba. In a similar, if less thorough, fashion I de-centre the mainstream model of (neo)liberalism by describing theorizations of value, materialities and moralities that situate this 'project of scale-making' (Tsing 2000: 347) in space and time.

The ethnographic chapters (chapters 3–6) associate these dominant models of state and market to the power geometries that emplace Tutaños unevenly in relation to such discourses and their spatialities. Like Katherine Neilson Rankin's (2004) account of honour in a Nepalese market village, I show how people differentially positioned in the Cuban political economic system incorporate, contest or otherwise manipulate the cultural politics of the nation state in order to manoeuvre between market and non-market spaces, and how some cultural-economic practices, in turn, (re)create inequalities. In line with other work in moral geography, I am interested in changing 'landscapes of care'[22] in rural Cuba, particularly how people with various relations to inside and outside forces live through scalar shifts in economic responsibility while continuing to 'domesticate'[23] dominant normative frameworks. For instance, I argue that while certain categories of persons, like Afro-Cubans, generally do not have access to remittances in hard currency or help from family members living *afuera* (few Afro-Cubans have enough money and/or political capital to leave Cuba), many Cubans of European decent do have such 'scale capabilities' (Swyngedouw 1997, cited in Rankin 2004: 65), which allow for more amenable permutations between state and market spaces.

In chapters 3 and 4, I use local narratives to reveal how consumers in Tuta position themselves in relation to the national collective. While Tutaños are increasingly uneasy about growing inequalities and scarcities, many continue to illustrate their moral commitment to revolutionary ideas and values. Chapter 3 centres on how the contradiction outlined in the previous chapter – between food as a public good in the state system and food as a private commodity in the market – affects the way ordinary people justify scarcities in Cuba. I start by briefly explaining actual divisions in food accessibility, particularly in terms of the distinction between an ideal model of national redistribution and more localized determinations of value and desire. A limited amount of basic necessities are provided to underprivileged Cubans through the Acopio (redistributive) system and at state-subsidized agricultural markets in pesos. Higher-quality imports and domestic products, some also produced by Cuba's growing population of small farmers, are sold either at higher-priced agricultural markets, through petty traders at world market prices or in hard currency at domestic and tourist shops (also at world market prices). The rest of chapter 3 shows how Tutaños deal discursively with scarcities and inequalities of access. I argue that local narratives of consumption draw from the scalar politics of the nation state, evident for example in Tutaños' particular use of irony. Local jokes about scarcity, which set the norms of the state against the difficult realities of post-1990s Cuba, do not entirely overturn socialist moralities relating person to community; indeed in some respects they maintain them. 'Lay normativities' (Sayer 2004) were also apparent in differentiations Tutaños made between what they wanted and what they 'liked', only using the latter in my presence so as to not seem 'interested' in receiving material gifts from me (a partial outsider with access to hard currency). In line with the oft-quoted adage of José Martí, 'The wine is sour, but it is *our* wine', I argue that narratives of scarcity and inequality in Tuta are tied to revolutionary ideas of the national collective.

Chapter 4 begins with a brief anthropological comparison, which reveals how values of redistribution and reciprocity coexist in Cuba as in other cultural contexts. I then argue that Tutaños create local counterparts to the state's version of communist redistribution and socialist reciprocity through consumer narratives of nourishment and hunger. Tutaños used the former word (*alimenta*) to indicate their two-way relationship with the state, which retains its distant role as provider for all those who sacrifice for the nation ('They must nourish us'). The word 'hunger' (*pasar hambre*) signifies just the opposite: a feeling of abandonment and redistributive injustice, which has been especially acute in the post-1990s period of scarcity ('They are letting us go hungry!'). Since the early 1990s, Tutaños have likely 'gone hungry' more than in any other period in revolutionary history. The present Cuban government claims that sacrifices are spread evenly and justly across the population and that collective benefits will accrue from market endeavours like tourism and taxes on private entrepreneurs. Indeed, as I argue, Raul Castro's economic reforms of 2011 do not overturn the communist redistribution of use values. Still, recent policies introduce new contradictions, as official discourse shifts from universal redistribution to a kind of reciprocity premised on what Young (2011: 11) calls a 'liability' or 'blame' model of economic responsibility.

Chapters 5 and 6 reveal how the moral economy in Cuba is shifting from the nation to neighbourhoods, households and individuals as Cuba opens its economy to domestic markets, tourism and privatized agriculture. On the ground, it is increasingly evident that market openings and continuities of political capital allow some people more access to the world of commodities than others. In chapter 5, I explain how inequalities allow for a scalar shift in communities of distribution and exchange, from the nation to the locality, though many people continue to justify appropriate economic practice in terms of the national moral economy. Political capital, tourism and the increasing number of workers employed '*por cuenta propia*' (literally, on one's own account; also called *particulares*) have escalated inequalities, redirecting the power of allocation from the state to individuals and households. At the same time, revolutionary ideas of 'culture', 'protection' and levels of 'interest' are used to define and evaluate people differentially positioned within systems of power and privilege. Drawing from themes outlined in previous chapters, I explain how Tutaños determine culturally appropriate exchanges between *jefes* and their workers, *particulares* and their clients and foreigners (like myself) and ordinary Cubans. While yielding to the central power of the PCC, acceptable exchanges often stray into illegal spaces for provisioning. They are made illegally licit, or justified in social terms, if they conform to shared understandings of how far one should cross the line from common interests to instrumentality.

In the penultimate chapter, I ask whether the national moral economy is applicable to a sector that has long been the most questionable in the socialist system: small farmers. Over the past few decades, the official idea of the small 'private' farmer has shifted from a short-lived egoist who works for his or her own benefit (a Leninist view), to the preserver and producer of sovereign property (land and food, respectively) and the forerunner of Cuba's agroecology movement. Though small farmers are in an economically advantageous position compared to most others in Cuban society, an official shift away from the industrial model occurred in the 1990s when the state initiated a political drive to cut imports of food and agrochemicals by distributing land in usufruct (use rights) to 'worthy' people. Like the categories of privileged people introduced in chapter 5, I argue that small farmers are both officially and socially evaluated according to how much they give back to their national and local communities. Formal controls include the creation of an alternative monetary system and special stores for small farmers, which prevent uneven access to luxuries otherwise available at hard currency stores. Social requirements include sharing produce, information and state-distributed bio-fertilizers. Despite such informal and formal rules for food production, it is clear that small farmers enter into market spaces, including the tourist market. Discrepancies between a national politics of redistribution and sustainability and a local politics of appropriate production and exchange are evident from my interviews, though, like the consumers discussed in chapters 3 and 4 and the traders discussed in chapter 5, farmers still justify their economic practices in terms of national values like collective property.

In the concluding chapter, I return to theoretical issues introduced in the first few chapters, again broadening the notion of 'moral economy' to incorporate both communitarian and liberal versions of community. I argue that moral economic projects to establish alternatives may be no better than their mainstream counterparts if they fail to 'jump scale' from top-down ideas of justice to the various scales in which everyday experiences are played out. Whether and how 'alternative' projects resist or counteract mainstream trends therefore depends on how successful they are at shifting from individual or local perspectives to wider normative and political economic relations, and back. I argue that projects to establish alternatives to mainstream economic

Figure 1.4   Two men walk down the road in Tuta. Source: author.

relations necessitate the creation of social collectivities with shared norms of environmental, social and economic justice. In this sense, the Cuban alternative is durable, for many Cubans still demonstrate a moral commitment to long-term ideas of national solidarity. This makes alternative provisioning systems in Cuba more formidable, if less extensive, than similar projects elsewhere. Relating theories of justice to observed practice, collected over an extended period of time, I aim to shift focus from transient prescriptions for Cuban transition to timeless concerns with redistributive justice, (in)equality, and the scope of community.

Notes

1   'Tuta' is a municipality of about 350 km² with a population density of 141 inhabitants/km². Its primary agricultural products are citrus, sugar cane and dairy, though increasingly small-scale farming is also practised. At least during the period of fieldwork, there were no 'organoponicos' or state-owned organic gardens in the municipality. Tuta is divided into five *consejos populares* (juridical councils or what I call towns), the largest of which bears the same name. As with personal names, I have disguised the name of this town where a majority of my informants live. Given the political sensitivity of some of the following material – particularly within Cuba – I do not provide details of its location in visual form.

2   A *paladar* is a small 'family-run' restaurant in Cuba, legally permissible only if the owners pay high taxes (usually in hard currency) to the Cuban state. Passed in June 1994, Resolution No. 4 prohibited non-family members to work in *paladares*, though 'ordinary' people as well as officials turned a blind eye to such legal controls. In April 2011, the Cuban government legalized the employment of

non-family members in *paladares* and other small businesses for the first time since the 1960s. The shift between legal and illegal (or formal and informal) economic activity is a central subtheme of this book, and it continues despite recent openings of the economy.

3   All salary estimates are based on my latest period of ethnographic fieldwork in 2011. It is possible that the basic official salary has increased for some Cubans since then, given further legalizations of the private (service) sector; however, as indicated in chapter 4, all legal entrepreneurs must pay hefty taxes to the state.

4   I first met Clare and Jorge (pseudonyms) – my 'parents' – in 2002 and developed a long-term relationship that led to our co-habitation during the fieldwork period. Our fictitious kinship is and was as cultural and emotional as practical, for as I explain later, I obtained official permission to live in their household as a member of the family.

5   References to 'Miami' or '*afuera* (outside)' encompass a wide variety of places where Cubans have emigrated, though most do live in Miami.

6   Mansvelt (2005: 96).

7   Cited by Dumont (1977: 108).

8   Hart (1986: 645).

9   Weiner (1992).

10   Holbraad (2000: 9).

11   In January 2013, Raul Castro eliminated the requirement that Cubans purchase an exit visa, an expensive and complicated system that prevented many so-called dissidents from leaving Cuba. Now a Cuban citizen may travel for up to 2 years abroad with their Cuban passport and an entry visa from the intended country of destination. The opening of foreign travel does not necessarily mean a mass exodus, however, since the average salary of 20–40 CUCs/month is not even enough to pay the increased fee for a Cuban passport (100 CUCs).

12   Fraser uses the term 'Westphalian' or 'Keynesian-Westphalian' to refer to the post-war inclination among 'first world' governments to treat economic rights and controls as matters internal to individual states. One example she gives (Fraser 2005: 1) is the Bretton Woods system, which separated out national economies for the purpose of maintaining functional international economic relations. Her idea of 'Westphalian' originates from the treaties of Westphalia of 1648 (also called the Peace of Westphalia), which ended the Thirty Years' War (1618–1648) by ratifying the sovereignty of individual European states.

13   That is, Holgado Fernández (2000); Berg (2004: 48); Roland (2006: 156); Pertierra (2007: 3, 40–1, 2011: 75–106); Settle (2007: 15).

14   The Tiv (or Tivi) are an ethno-linguistic group of people of West Africa (Nigeria and Cameroon).

15   In his preface to Paul Bohannan and George Dalton's classic volume, *Markets in Africa* (1962), Melville Herskovits argued against the kind of analysis which poses 'the market' as a homogenous social fact that applies to all societies in the same way. Instead, he claimed that societies with combinations of barter, 'money-barter' (or scrip) and money are different 'in kind' (rather than in degree) from societies with an all-encompassing market, in which economic transactions are carried out using a single 'general purpose money' (Herskovits 1962: 16). The formalist/substantivist debate that ensued in anthropology illustrated that

Bohannan and Dalton's position was flawed as it was still based on the primitive/modern polemic. Economic anthropologists now use Bohannan and Dalton's ideas and those of earlier substantivists (especially Karl Polanyi) to show that even modern economies cannot be entirely explained by the market principle.

16 That is, Long and Richardson (1978); Gregory (1982, 1997); Hart (1986); Parry and Bloch (1989); Dilley (1992); Miller (1997, 2001); Graeber (2001); Gudeman (2008).

17 For instance, Wendy James, retired Professor of Social Anthropology and Emeritus Fellow of the Institute of Social and Cultural Anthropology, Oxford, writes: 'Danny Miller has argued that the idea of consumption itself, with all the house decoration, food preparation, and shopping that goes into it, is a kind of ritual, imbued with morality and symbolism, even religious overtones, though some may still feel the need for an anchor in the basic facts of capitalism and the way it tends to promote inequalities. The consumer is king as long as he – or more usually she – has a full purse, but not everyone involved in the production of the goods we see in the supermarket has this power' (James 2003: 257).

18 For an insightful discussion of the ways top-down norms really 'matter' to people see Sayer (2011).

19 Following earlier anti-imperialist revolutionaries of the region, particularly the Venezuelan, Simón Bolívar (1783–1830), Cuban leaders have long had visions for a socialist form of Latin American (and Caribbean) unity. For some at least, this dream is approaching reality as countries like Venezuela, Ecuador, Brazil, Dominica, St Vincent and the Grenadines, Antigua and Barbuda, Nicaragua and Bolivia continue to develop the economic, social and political trade bloc appropriately named the Bolivarian Alliance for the (Peoples of Our) America (*Alianza Bolivariana para los Pueblos de Nuestra América*, ALBA). Formed by the late President Hugo Chavez of Venezuela in 2004, this emerging trade bloc is something to watch, particularly as ALBA has plans to introduce a new currency called the Sucre, which would allow for more autonomous trade and barter between member countries.

20 At least during the fieldwork period, the transport of foodstuffs in bulk, especially goods monopolized by the state such as coffee, was prohibited on the island. In 2007, penalties for this offence were as severe as those for drug smuggling in other countries.

21 Remembering is also key in politically charged contexts, such as Tuta, where interviewees do not feel comfortable being recorded.

22 Read and Thelen (2007), cited in Stenning *et al.* (2010: 175).

23 Stenning *et al.* (2010), in reference to Creed (1998).

References

Abrams, Philip. (1988) Notes on the difficulty of studying the state, *Journal of Historical Sociology* 1(1): 58–89.

Appadurai, Arjun. (1986) Introduction. In *The Social Life of Things: Commodities in Cultural Perspective*, edited by Arjun Appadurai. Cambridge: Cambridge University Press, pp. 3–63.

Barnett, Clive, Paul Cloke, Nick Clarke and Alice Malpass (eds). (2011) *Globalizing Responsibility: the Political Rationalities of Ethical Consumption*. Oxford: Wiley Blackwell.

Berg, Mette Louise. (2004) Tourism and the revolutionary new man: the specter of *jineterismo* in late 'Special Period' Cuba, *Focaal: European Journal of Anthropology* 43: 46–55.

Bohannan, Paul and Laura Bohannan. (1968) *Tiv Economy*. Evanston: Northwestern University Press.

Bohannan, Paul and George Dalton. (1962) Introduction. In *Markets in Africa*, edited by Paul Bohannan and George Dalton. Evanston: Northwestern University Press, pp. 1–26.

Boltanski, Luc and Laurent Thévenot. (2006) *On Justification: Economies of Worth* (transl. by Catherine Porter). Princeton: Princeton University Press.

Callon, Michel. (2005) Why virtualism paves the way to political impotence: a reply to Daniel Miller's critique of *The Laws of the Markets*, *Economic Sociology* 6(2): 3–20.

Callon, Michel and Bruno Latour. (1981) Unscrewing the big leviathan: how actors macro-structure reality and how sociologists help them to do so. In *Advances in Social Theory and Methodology: Towards an Integration of Micro and Macro-Sociology*, edited by K. Knorr-Cetina and A.V. Cicouvel. London: Routledge, pp. 277–303.

Carrier, James and Daniel Miller. (1998) *Virtualism: a New Political Economy*. Oxford: Berg.

Castro Ruz, Raúl. (2010) Discurso pronunciado por el general de ejército Raúl Castro Ruz, Presidente de los Consejos de Estado de Ministros, en el quinto period ordinario de sesiones de la VII Legislatura de la Asamblea Nacional del Poder Popular. Palacio de Convenciónes, 1 Aug.

Cohen, Anthony P. (2000) Introduction: discriminating relations – identity, boundary and authenticity. In *Signifying Identities: Anthropological Perspectives on Boundaries and Contested Values*, edited by Anthony P. Cohen. London: Routledge, pp. 1–14.

Cook, Ian, Kersty Hobson, Lucius Hallett IV *et al.* (2010) Geographies of food: afters, *Progress in Human Geography* 35(1): 104–20.

Cox, Kevin. (1998) Spaces of dependence, spaces of engagement and the politics of scale, or: looking for local politics, *Political Geography* 17: 1–23.

Crang, Philip, Claire Dwyer and Peter Jackson. (2003) Transnationalism and the spaces of commodity culture, *Progress in Human Geography* 27(4): 438–56.

Creed, Gerald. (1998) *Domesticating Revolution: From Socialist Reform to Ambivalent Transition in a Bulgarian Village*. University Park: Penn State University Press.

Dilley, Roy. (1992) Introduction. In *Contesting Markets: Analogies of Ideology, Discourse and Practice*, edited by Roy Dilley. Edinburgh: Edinburgh University Press, pp. 1–26.

Dresch, Paul. (1976) Economy and ideology: an obstacle in materialist analysis, *Journal of the Anthropological Society of Oxford* 3(2): 55–77.

Dumont, Louis. (1977) *From Mandeville to Marx: the Genesis and Triumph of Economic Ideology*. Chicago: University of Chicago Press.

Dumont, Louis. (1986 [1983]) *Essays on Individualism: Modern Ideology in Anthropological Perspective*. Chicago: University of Chicago Press.

Fabienke, Rikke. (2001) Labor markets and income distribution during the crisis and reform. In *Globalization and Third World Socialism: Cuba and Vietnam*, edited by Claes Brundenius and John Weeks. London: Palgrave, pp. 102–28.

Fine, Ben. (2002) *The World of Consumption. The Material and Cultural Revisited*. London: Routledge.

Fine, Ben and Ellen Leopold. (1993) *The World of Consumption*. London: Routledge.

Fraser, Nancy. (2009) *Scales of Justice: Reimagining Political Space in a Globalizing World*. New York: Colombia University Press.

Fraser, Nancy. (2005) Reframing justice in a globalizing world, *New Left Review* 36 (Nov–Dec): 69–88.

Fuller, Duncan and Andrew E.G. Jonas. (2003) *Alternative Economic Spaces*, edited by Andrew Leyshon, Roger Lee and Colin Williams. London: Sage Publications.

Fuller, Duncan, Andrew E.G. Jonas and Roger Lee. (2010) *Interrogating Alterity: Alternative Economic and Political Spaces*. London: Ashgate.

Geertz, Clifford. (1973) *The Interpretation of Cultures: Selected Essays*. New York: Basic.

Gibson-Graham, J.K. (1996) *The End of Capitalism (as We Knew it): a Feminist Critique of Political Economy*. Minneapolis: University of Minnesota Press.

González Maicas, Zoila. (2012) Challenges and transformation in the Cuban economy, paper presented at the University of the West Indies, 20 Mar.

Graeber, David. (2001) *Toward an Anthropological Theory of Value: the False Coin of Our Own Dreams*. New York: Palgrave.

Gregory, Chris. (1982) *Gifts and Commodities*. London: Academic Press.

Gregory, Chris. (1997) *Savage Money: the Anthropology and Politics of Commodity Exchange*. London: Harwood.

Gudeman, Stephen. (2008) *Economy's Tension: the Dialectics of Community and Market*. New York: Berghahn Books.

Hart, Keith. (1986) Heads or tails? Two sides of the coin, *Man* 21(4): 637–58.

Herskovits, Melville. (1962) Preface. In *Markets in Africa*, edited by Paul Bohannan and George Dalton. Chicago: Northwestern University Press, vii–xvi.

Herzfeld, Michael. (2005) *Cultural Intimacy: Social Poetics in the Nation State* (2nd revised edition). New York: Routledge.

Holbraad, Martin. (2000) *Money and Need: Havana in the Special Period, Presentation to the Annual Post-Socialism Workshop*, University College, London.

Holbraad, Martin. (2004) Response to Bruno Latour 'Thou shall not freeze frame', *Mana. Estudios de Antropología Social* 10(2): 349–76.

Holgado Fernández, Isabel. (2000) *¡No es Facil! Mujeres Cubanas y la Crisis Revolucionaria*. Barcelona: Icaria.

Humphrey, Caroline. (1989) 'Janus-faced signs' – the political language of a Soviet minority before Glasnost. In *Social Anthropology and the Politics of Language*, edited by Ralph Grillo. London: Routledge.

James, Wendy. (2003) *The Ceremonial Animal: a New Portrait of Anthropology*. Oxford: Oxford University Press.

Jenkins, Timothy. (1994) Fieldwork and the perception of everyday life, *Man* (N.S.) 29(2): 433–55.

Jonas, Andrew. (2010) 'Alternative' this, 'alternative' that … : Interrogating alterity and diversity. In *Interrogating Alterity Alternative Economic and Political Spaces*, edited by Duncan Fuller, Andrew Jonas and Roger Lee. London: Ashgate, pp. 3–30.

Kapcia, Antoni. (2000) *Cuba: Island of Dreams*. Oxford: Berg.

Lee, Roger. (2006) The ordinary economy: Tangled up in values and geography, *Transactions of the Institute of British Geographers* (N.S.) 31: 413–32.

Lee, Roger. (2010) Spiders, bees or architects? Imagination and the radical immanence of alternatives/diversity for political-economic geographies. In *Interrogating Alterity:Alternative Economic and Political Spaces*, edited by Duncan Fuller, Andrew Jonas and Roger Lee. London: Ashgate, pp. 273–88.

Leyshon, Andrew, Roger Lee and Colin C. Williams (eds). (2003) *Alternative Economic Spaces*. London: Sage Publications.

Long, Norman and Paul Richardson. (1978) Informal sector, petty commodity production and the social relations of small-scale enterprise. In *The New Economic Anthropology*, edited by John Clammer. London: Macmillan, pp. 176–203.

Mansvelt, Juliana. (2005) *Geographies of Consumption*. London: Sage Publications.

Massey, Doreen. (1991) A global sense of place, *Marxism Today* 38: 24–9.

Massey, Doreen. (2004) Geographies of responsibility, *Geografiska Annaler* 86(1): 5–18.

Massey, Doreen. (2011 [2005]) *For Space*. London: Sage Publications.

Miller, Daniel. (1997) *Capitalism: an Ethnographic Approach*. Oxford: Berg.

Miller, Daniel. (2001) *The Dialectics of Shopping*. Chicago: University of Chicago Press.

Miller, Daniel. (2011) *Tales from Facebook*. Cambridge: Polity Press.

Myrdal, Gunnar. (1953). *The Political Element in the Development of Economic Theory*. London: Routledge and Kegan Paul.

Narotzky, Susana. (2005) Provisioning. In *A Handbook of Economic Anthropology*, edited by James Carrier. Cheltenham: Edward Elgar.

North, Peter. (2005) Scaling alternative economic practices? Some lessons from alternative currencies, *Transactions of the Institute of British Geographers* 30(2): 221–33.

Parry, Jonathan and Maurice Bloch. (1989) Introduction: money and the morality of exchange. In *Money and the Morality of Exchange*, edited by Jonathan Parry and Maurice Bloch. Cambridge: Cambridge University Press, pp. 1–32.

Peck, Jamie. (2010) *Constructions of Neoliberal Reason*. Oxford: Oxford University Press.

Pertierra, Anna Cristina. (2007) Cuba: the struggle for consumption, doctoral thesis, University College London.

Pertierra, Anna Cristina. (2011) *Cuba: the Struggle for Consumption*. Coconut Creek: Caribbean Studies Press.

Polanyi, Karl. (1944) *The Great Transformation: the Political and Economic Origins of Our Time*. Boston: Beacon Press.

Rankin, Katherine Neilson. (2004) *The Cultural Politics of Markets: Economic Liberalization and Social Change in Nepal*. London: Pluto Press.

Richards, Audrey. (1932) *Hunger and Work in a Savage Tribe: a Functional Study of Nutrition among the Southern Bantu*. London: George Routledge and Sons.

Roitman, Janet. (2005) *Fiscal Disobedience: an Anthropology of Economic Regulation in Central Africa*. Princeton: Princeton University Press.

Roland, L. Kaifa. (2006) Tourism and the *negrificatión* of Cuban identity, *Transforming Anthropology* 14(2): 151–62.

Samers, Michael and Jane Pollard. (2010) Alterity's geographies: socio-territoriality and difference in Islamic banking and finance. In *Interrogating Alterity: Alternative*

Economic and Political Spaces, edited by Duncan Fuller, Andrew Jonas and Roger Lee. London: Ashgate, pp. 47–58.

Sayer, Andrew. (2001) For a critical cultural political economy, *Antipode* 33(4): 687–708.

Sayer, Andrew. (2004) Restoring the moral dimension: acknowledging lay normativity. http://www.lancs.ac.uk/fass/sociology/research/publications/papers/sayer-restoring-moral-dimension.pdf (last accessed 16 May 2013).

Sayer, Andrew. (2011) *Why Things Matter to People: Social Science, Values and Ethical Life*. Cambridge: Cambridge University Press.

Settle, Heather. (2007) On giving and getting by in Special Period Cuba: love, belief and survival in the world of the permanent crisis, doctoral thesis, Duke University.

Smith, Adrian. (2002) Culture/economy and spaces of economic practice: positioning households in post-communism, *Transactions of the Institute of British Geographers* (N.S.) 27: 232–50.

Smith, Neil. (1984) *Uneven Development, Nature, Capital and the Production of Space*. Oxford: Basil Blackwell.

Stenning, Alison, Adrian Smith, Alena Rochovská and Dariusz Świątek. (2010) *Domesticating Neo-liberalism: Spaces of Economic Practice and Social Reproduction in Post-Socialist Societies*. Oxford: Wiley Blackwell.

Suárez Salazar, Luis. (2000) *El Siglo XXI: Posibilidades y Desafíos para la Revolución Cubana*. Havana: Editorial de Ciencias Sociales.

Tsing, Anna. (2000) The global situation, *Cultural Anthropology* 15(3): 327–60.

Watts, Michael. (1999) Commodities. In *Introducing Human Geographies*, edited by Paul Cloke, Philip Crang and Mark Goodwin. Oxford: Oxford University Press.

Weiner, Annette B. (1992) *Inalienable Possessions: the Paradox of Keeping-While-Giving*. Berkeley: University of California Press.

Whitehead, Laurence. (2007) On Cuban political exceptionalism. In *Debating Cuban Exceptionalism*, edited by Bert Hoffman and Laurence Whitehead. New York: Palgrave Macmillan, pp. 1–26.

Williams, Colin C. (2005) *A Commodified World? Mapping the Limits of Capitalism*. London: Zed Books.

Wilson, Marisa. (2012) The moral economy of food provisioning in Cuba, *Food, Culture and Society* 15(2): 277–91.

Wolf, Eric. (1999) *Envisioning Power: Ideologies of Dominance and Crisis*. Berkeley: University of California Press.

Young, Iris Marion. (1990) *Justice and the Politics of Difference*. Princeton: Princeton University Press.

Young, Iris Marion. (2011) *Responsibility for Justice*. Oxford: Oxford University Press.

Chapter Two
The Historical Emergence of a National Leviathan

Geography dictates that, eventually, Cuba will become fully integrated not only into the global economy but also into the Caribbean Basin. In preparing to take advantage of the widening of the Panama Canal by constructing shipping facilities at Mariel to service the Eastern seaboard of the United States, and by authorizing new golf courses and boat marinas that only American citizens could fully occupy, the Cuban authorities are signaling that they understand the powerful gravitational pull of geography – that Cuba and the United States will, inevitably, once again become economic partners. In approaching Cuban economic reform, the United States should join with the international development community in nudging forward that irresistible flow of history. (Richard Feinberg 2011: 100–1)

In 1984, Ronald Reagan introduced the Caribbean Basin Initiative, with the dual aim of quelling communist tendencies in the region and integrating Latin America and the Caribbean into their 'natural' market (Suárez Salazar 2000: 138): the United States. The Caribbean Basin Initiative was a precursor to current proposals for a Free Trade Area of the Americas and to existing trade agreements like the North American Free Trade Agreement. Such agendas are moral and normative as well as political and economic; they are based on a uniform spatio-temporal vision that places all national economies on 'the' global trajectory towards commodification: 'While the West is viewed as "advanced" due to the *supposed* extent to which commodification has permeated its economic practices, both the terms "transition" and "developing" economies construct these nations as becoming

Everyday Moral Economies: Food, Politics and Scale in Cuba, First Edition. Marisa Wilson.
© 2014 John Wiley & Sons, Ltd. Published 2014 by John Wiley & Sons, Ltd.

commodified' (Williams 2005: 155; my emphasis). In this chapter, I show how political economic visions of Cuba, such as the one cited above, recall pre-1959 statements that pose economic ties between the United States and Cuba as geographically natural and historically inevitable but also as morally right and good. As Peter Hulme (2012: 354–5)[1] has recently argued, Cuba has long been metaphorized as a 'key' to the region, a spatio-temporal imaginary that has, since the 16th century, represented dominant interests in both countries. Cuba as 'key' (to a future liberal or socialist hemisphere) is also a moral economic vision of what Cuba 'ought' to be.

While the term 'moral economy' is usually applied to collective reactions *against* the anonymity of markets, (neo)liberalism is as moral an economy as any other. Indeed, if all moral economies 'embod[y] norms and sentiments regarding the responsibilities and rights of individuals and institutions with respect to others' and go 'beyond matters of justice and equality, to conceptions of the good' (Sayer 2000: 79), then neoliberalism is simply an (admittedly multifaceted) moral economy that favours the rights of consumers to satisfy their individual preferences, and the responsibilities of both consumers and institutions (e.g. states) to enable the free flow of supply and demand. As Luc Boltanski and Laurent Thévenot (2006) argue, neoliberalism is tied to a higher 'common good': a moral contract between individuals who implicitly agree to the universality of commodified futures:

> The conception of the individual required by economists to make their argument imposes constraints on the social actor that make him a moral being. … [M]oral capacity is presupposed in the construction of an order of market exchanges among persons, who must be capable of distancing themselves from their own particularities in order to reach agreement about external goods. (Boltanski and Thévenot 2006: 27)

Liberalism once freed economically ambitious individuals from political hierarchies imposed by monarchies and states, but it instituted its own hierarchy in the process (Dumont 1986 [1983]), disguised under a pact between 'free' economic individuals. According to this hierarchy, the 'worthy' (Boltanski and Thévenot 2006) are economically rational; 'unworthy' people do not make rational decisions in the market and so are not considered fully human. 'Since the individual is human only in so far as free, and free only in so far as a proprietor of himself, human society can only be a series of relations between sole proprietors, i.e. a series of market relations' (Macpherson 1979 [1962]: 264). Underlying the moral economy of (neo)-liberalism is a supreme value – the 'just' price – achieved when all economically rational individuals establish equilibrium between supply and demand through their unfettered choice of which commodities to buy in the market.

Richard Feinberg (cited above),[2] draws from the neoliberal moral economy to defend Cuba's 'inevitable' integration into the Caribbean Basin.

In his highly influential report of 2011, Feinberg claims that Cuba will inexorably rejoin[3] the International Monetary Fund (IMF), a primary actor that drives neoliberalization. According to the prevailing prediction, if Cuba rejoined the IMF the latter would encourage the elimination of price subsidies for basic foodstuffs and other public goods, the argument being that this would free up prices and motivate market competition among Cuba's small-scale entrepreneurs (Davis 2006). Like earlier liberal models, this prediction relies on the assumption that freeing up individuals' 'natural' proclivity to satisfy their own interests ensures the general welfare of all:

> The argument runs roughly as follows. We live in society and depend on each others' services. ... These services are rendered most efficiently if we allow free play to self-interest. Acquisitiveness is a force which Providence has planted into our nature. Its fruits accrue to the benefit of all, if only we let it move without hindrance. Whenever someone increases his income, all benefit. For he can only succeed by offering to his fellows better and cheaper services than his rivals; hence consumption guides and directs production. (Myrdal 2002 [1953]: 44–5)

Feinberg poses Cuba's entry into the IMF as both universally beneficial and inevitable. Assuming that Cuba will re-enter the US 'sphere of influence', he emphasizes Cuba–US relations over other spatial and economic relations based on different values, like so-called 'post-neoliberal'[4] barter terms of trade already established between Cuba and Venezuela, Ecuador and other Latin American and Caribbean countries (through the Bolivarian Alliance for the Peoples of Our America, ALBA) or like 'solidarity' ties with countries in Latin America, Europe, Africa, Asia and the Middle East. Alternative futures are also discounted in Feinberg's account. Instead of recognizing a 'simultaneity of stories so far' (Massey 2011 [2005]: 5), he relies on a universal projection of 'economic progress' (Feinberg 2011: 3): a good example of what Colin C. Williams (2005) calls the commodification thesis.

This way of 'framing inevitability' (Massey 2011 [2005]: 85) is directly opposed to Cuban socialism, which has come to be associated with another spatio-temporal vision of progress that emerged from a history of revolutionary nationalism. In contrast to Feinberg's cultural politics of neoliberalism, the politics of Cuban identity involves the long-term commitment to ensure collective entitlements to food and other necessities through the moral duty of producers (rather than consumers), an alternative value system based not on acquisitiveness, but on hard work and self-sacrifice. In theory, the patriotic and socialist commitment of Cuban citizens is incommensurable with market prices and preferences: 'Preferences allow for substitution, commitments do not, and some may not be in our self-interest' (Sayer 2011: 126–7). In practice, we shall see that *both* prices *and* patriotism influence everyday moral economies in Cuba.

Understanding historical and geographical contingencies that have led to the contrast between two moral and political economic projects of scale-making – that of the United States and Cuba – opens up spaces for other 'stories so far' that emerge in-between (or outside of) state and market spaces. In Cuban history as well as everyday life, 'ordinary' economic logics and actions complicate geopolitical and normative oppositions between social commitments and individual preferences. This chapter provides historical examples of such incongruities, highlighting some opposing ways Cuban nationalism has been constructed by Cubans within and outside of national borders and drawing from pre-revolutionary values of commodities and culture, work and reciprocity. In subsequent chapters, I continue to show how people in Tuta switch from one value system to another, though I argue that everyday moral economies are more closely connected to long-term revolutionary values than to (neo)liberal values of individualism.

In comparing two moral economic visions (what I have called Leviathans), each with their own 'space-time' (Castree 2009), I am not simply setting up 'straw men' with which the leftist analyst may console herself (Barnett 2005) and then deconstruct or prove as false (Castree 2006). Rather, I take the power of moral economic discourses seriously, and contend that ethnographic and historical research is useful for uncovering how general projects for economic change come to shape everyday moral economic life in particular places and at particular times. Since the primary goal of anthropology is to find human generals in human particulars, its ethnographic and historico-comparative method may open up new avenues for emerging geographies of neoliberalization.[5] Providing ethnographic evidence that reveals relations and contradictions between economic discourse and everyday knowledges and practices at least partially resolves the dilemma of the geographer who wants to 'have her cake and eat it too' by emphasizing *both* political economic hegemony *and* geographico-historical difference (Castree 2006).

As for Cuban socialism so I surmise for neoliberalism (I can only surmise here, for this book is really about Cuban socialism),[6] the abstract inhabits the substantive when 'it' is concretized into everyday justifications for what people 'ought' to do: 'I ought to make myself feel better with some retail therapy!' While recognizing that there are multiple neoliberalisms (Larner 2003; Leitner and Miller 2007; Peck 2010) and socialisms in real 'time-space', we should also take the creative formation of uniform political and moral economic frameworks seriously, for they become the 'ghosts' (Tsing 2000: 351) that haunt everyday moralities and materialities. It is my task in the ethnographic chapters to prove the presence of such ghosts through the lens of everyday food provisioning (see chapters 3–6). In the current chapter, I use comparative and historical analysis to uncover how the dominant moral economy in Cuba developed as a discursive counterpart to its foil, (neo)liberal capitalism, in the first place.

The chapter is divided into five parts. In the first and second, I outline some historical contours of Cuban nationalism by highlighting common revolutionary values and mythical figures like José Martí, nationalist frameworks used differently according to various interests at play within and outside of Cuba. The third part shifts emphasis to localized values, drawing from life histories to address continuous if contradictory values of work, commodities, reciprocity and culture in Tuta. In the fourth part, I introduce a second mythical figure in Cuba – Ernesto 'Che' Guevara – and illustrate the importance of his economic policies for framing the national moral economy in Cuba. The final part reveals the continued moral and political economic relevance of Guevarian and Martían thought in present-day Cuba.

The first revolution 'of the humble, for the humble and by the humble'[7]

> There are those who selfishly confuse patriotism with the price of sugar ...
> (Portell Vilá 1995 [1939]: lxvii)

The discourse of a unified revolutionary identity in Cuba has long been 'mediated by its perception of the identity of the other' (Cohen 2000a: 9). Cuban nationalists in both pre- and post-Castro Cuba used the binary between 'us' and 'them' strategically, as nationalists in other places are wont to do (Herzfeld 2005: 15). By the time Fidel Castro came to power, a Manichaean discourse had emerged, which posed the idea of a national Fight or *Lucha* for social justice and universal access to collective assets (such as land and food) against the corruption and greed of Cuban elites and US businessmen and politicians (i.e. 'those who selfishly confuse patriotism with the price of sugar'). As we shall see, this sense of historical injustice, also a common tool of nationalism (Cohen 2000b: 164), has been interpreted differently according to the differential positioning of Cubans living within and outside the country.

The central dichotomy of Cuban nationhood – between a collective *Lucha*, on the one hand, and mercenary individuals, on the other – was established in Cuba long before Castro's 'triumph' in 1959. During the second half of the 19th century, the idea of the Cuban *patria* (literally 'homeland', but synonymous with 'nation' since the 1700s; Kapcia 2000: 22–3) was divided by two factions: loyalists to Spain, who were mostly merchants and landowners residing in Havana, and *independentistas* (pro-independence fighters), whose origins were in rural areas of the east. The split was partly defined in terms of economic status – those who benefited and those who suffered from the sugar industry – but also in terms of the rural–urban divide. In fact, the first rebel army to use the term *'Cuba Libre'*

was largely comprised of Afro-Cubans residing in the rural east (Oriente).
Their rural and humble – if not racial – origins became a primary 'code'
used by later revolutionaries to defend national sovereignty (Kapcia
2008: 10–11):

> [W]hen the capital city is seen by proto-nationalist groups as the place where
> their interests are being betrayed, then other places, other regions, can become
> metonyms for the post-colonial state, a role that Oriente has increasingly
> played in Cuba over the last 150 years because of it being the setting for
> almost all revolutionary initiatives. In that sense Oriente may be thought of as
> the *heart* of revolutionary Cuba. (Hulme 2012: 353)

In contrast to the Ten Years' War (1868–1898), the War of Independence
(1895–1898) was led by poor blacks as well as whites, by peasants as well as
workers (Pérez 1995 [1988]: 160): those with the least to gain from economic
relations with urban elites (some of whom joined the rebel cause) or powerful
outsiders. As historian Louis Pérez Jr argues, the Cuban War of Independence
was 'a war waged against the beneficiaries of colonialism by the victims of colonialism' (Pérez 1995 [1988]: 162). Destruction of sugar lands by arson was a
primary strategy deployed by the *mambisas* (former slave groups who became
rebel soldiers), and the tactic left peasants as well as middle and upper class
landowners in a state of bankruptcy. In 1895, the Dominican Máximo Gómez,
Capitan General of the rebel army, ordered a moratorium on all economic
activity in the countryside, setting fire to all properties with a plan to redistrib-
ute them during peacetime (Pérez 1995 [1988]: 163). Spanish General Weyler
reacted with a 'war with war' policy, forcibly relocating rural dwellers to Havana
to prevent smallholders from collaborating with the *mambisa* soldiers.

The war and its aftermath eliminated an entire class of landowners.
Lands lost during the 1895–1898 period would never be returned, for
the Platt Amendment (addressed below) was to guarantee only post-war
property relations. According to Pérez (1995 [1988]: 197–8), by 1905 US
citizens owned about 60% of rural property. The rest was divided between
first-generation Spanish (15%) and those who now identified themselves as
Cubans, who owned only 25%. With the planter class eliminated, the reins
of political and economic power were taken up by urban elites with strong
ties to foreign politics and markets:

> The Cuban upper class ... was incapable of developing an independent
> economic or political role. ... Unable to be an independent bourgeoisie, it
> was also unable to act as a *national* bourgeoisie. Many of its members were
> former Spanish or American nationals. Nor could they forge an effective tie
> to a landed creole aristocracy of the kind existent in the hinterland of other
> Latin American countries, since this group had been effectively replaced by
> corporate managers operating under US auspices. (Wolf 1969: 260; emphasis
> in original)

The national sugar industry had been in crisis even before the mass destruction and post-war reorganization of the Cuban countryside. In 1894, the Wilson Tariff curtailed sugar exports to the United States, now Cuba's most important market. Exports to the United States fell from 800 000 t in 1895 to 225 231 t in 1896 (Thomas 1998 [1993]: 24). The Wilson Tariff was the first of many policies to expose Cuba's dependency on the market whims of the United States. The resulting economic crisis sparked a renewed *mambisa* call for *Cuba Libre* in 1895 (Thomas 1998 [1993]: 24).

For many wealthier *habaneros* (inhabitants of Havana), however, the United States offered a different vision of *Cuba Libre*. Their path to freedom was made through access to consumer durables and luxuries from the United States, a market demand that would continue to grow in the 20th century. Cuban nationhood would come to be defined in terms of this polarity between urban consumers and political elites – the annexationists whom I will later call the 'pro-imperialist nationalists' (following Guerra 2005: 16) – and the more radical tradition memorialized by *mambisa* resistance in rural Cuba. But the much-touted dichotomy between the rural, humble revolutionary and the mercenary, pro-US urban elites did not capture the diversity of factions that supported the *mambisa* cause. While Gómez's army of *mambisas*, workers and peasants never made it to Havana, many *habaneros* – particularly intellectuals and military leaders – supported their struggle for independence. All were eager to see an end to Weyler's violent and economically crippling 'war with war' policy.

The different factions in the War of Independence were unified by a singular leader who was to become the key myth of Cuban national identity: the slight poet and revolutionary who met an early death, majestically and conspicuously on a white horse, in his first experience on the battlefield on 19 May 1895. In creating a single Partido Revolucionario Cubano (Cuban Revolutionary Party), José Martí (Figure 2.1)

> managed to fuse the movement's disparate nationalism and its growing radicalism into a vision of *Cuba Libre* that went beyond political independence to advocate a socially equal Cuba; moreover, he alone of all Cuban nationalist leaders was aware of, and increasingly feared, the growth of US power and of quasi-imperialist thinking within the United States, warning his fellow countrymen of the dangers to Cuban independence that might be posed by those designs. (Kapcia 2008: 13)

In the decades following Cuba's independence from Spain in 1898, the island experienced an 'unusually explicit' (Kapcia 2008: 7) neocolonialism under the political, legal and military control of the United States and US-led political elites. For many who witnessed the increasing encroachment of US markets in everyday life, there was a 'steady undermining of a sense of collective identity and self-belief and a growing frustration, resentment and anger at what was seen as the betrayal of the ideals of 1895, of Martí's memory, and of the unity and purpose

Figure 2.1   A statue of José Martí pointing at the US Interests Section in Havana, surrounded by black flags. Source: author.

of the independence struggles' (Kapcia 2008: 16). The popular sense of historical injustice was politicized after the sugar crisis of the 1920s, and particularly in the early 1930s when a new generation of revolutionary nationalists was to emerge who continued to rely on central symbols of Cuba's 'creation myth' (Guerra 2005: 7): *Lucha* and self-sacrifice, agrarianism and humility (Kapcia 2000).

This creation myth was based on a vision of the nation as divided between insiders and outsiders and, in particular, a moral outrage against outsiders within. Thus revolutionary nationalists of the 1930s and then the 1950s contrasted the series of mercenary and often corrupt presidents and officials who 'selfishly confuse[d] patriotism with the price of sugar ...' (Portell Vilá 1995 [1939]: lxvii) with Cubans who put the collective Fight against injustice and imperialism before their own interests in profits. As director of the Junta Cubana de Renovación Nacional (Cuban Committee for National Renewal), Fernando Ortiz wrote in the early 1930s: 'The Cuban people want to be free as much from the foreigners who abuse the flag as from the citizens who violate it and will end up burying it' (Pérez 1995 [1988]: 236).

José Martí and contradictions of Cuba's creation myth

[Myth] … is not concerned so much with a succession of events as with the moral significance of situations. (Evans-Pritchard 1962: 53)

We have seen that conflicting views of freedom and nation have existed in Cuba at least since the second half of the 19th century. Despite such dichotomies of nationhood, nearly all 20th-century political figures appealed to values of Cuba's creation myth. In no other figure do these values coalesce more powerfully than in José Martí. Martí was the first of two mythical figures to give the ongoing project for Cuban sovereignty a moral foundation. The second was Ernesto 'Che' Guevara, to whom I shall return. Along with Fidel Castro, Guevara is usually associated with transcendental values such as the collective Fight of 'morality, duty and self-sacrifice'. As Cuban historian Antoni Kapcia argues, however, 'these ideas, in fact, hav[e] a *martiano* pedigree' (Kapcia 1986: 64).

Martí's writings about the national Fight for social justice and associated values have been selectively appropriated by a range of Cuban groups and often used in very contradictory ways. Indeed, ideas and concepts used to substantiate myths are not fixed; signs such as *Lucha* are multi-accentual rather than uni-accentual (Humphrey 1989: 145–6). Members of the Cuban diaspora in Miami and other North American cities have used the *martiano* idea of *Lucha* in a way that directly contradicts the concept as appropriated later by pro-Castro revolutionaries. It is worth briefly returning to the present to consider this exile community, if only because Cubans in the United States (and, to a lesser degree, Europe) have played a crucial role in shaping Cuban national identities. Though the influence of this émigré community is largely absent in official discourse (Duany 2000: 17), Martí himself (along with other Cuban intellectuals) was in exile in either the United States or Europe for over half his life.

As with Cubans who continue to live on the island, the Miami diaspora follow Martí in identifying the *Lucha* with Cuba's struggle for independence against Spain in the late 19th century (De la Torre 2003: 30). And, like other nationalists in Cuba (i.e. the 'popular nationalists' and the 'revolutionary nationalists' (Guerra 2005: 16); I will consider these groups in turn), the Miami diaspora and earlier 'pro-imperialist nationalists' also saw José Martí as an apostle: a spiritual and mythical figure who would lead the nation into the modern world. For the pro-imperialist nationalists and their successors in Miami, however, modernity and progress would come with market freedoms and technological advancements enabled by relations with the United States. In claiming to lead Cuba towards economic progress, Cuban elites in the first half of the 20th century invoked the tenets of a unique émigré nationalism whose constant referencing of US interests and social values made it 'pro-imperialist' long before US imperialism asserted itself on Cuba's war-ravaged shores. ... [T]he nationalism these Cubans expressed was, from the beginning, dependent on the United States for its ideological legitimacy. Achieving modernity through gradual social change and rapid, foreign-financed technological improvement were two of this nationalism's primary goals. (Guerra 2005: 64)

For the pro-imperialist nationalists of the Cuban republic and many Cuban émigrés of the post-1959 period, the *Lucha* that Martí began was consistent with the North American model of modernity and progress. Indeed, the contemporary Miami diaspora sees their present Fight against the Castro brothers as a continuation of Martí's *Lucha* against the kind of corruption and backwardness that inhibited the island's economic progress since the time of Spanish colonialism (De la Torre 2003: 30). This version of Martí's Fight emerged in the context of a 'crisis of legitimacy' (Guerra 2005: 36) of the Cuban republic, particularly the use of public funds (especially the lottery) for private benefit. Responding in typical populist fashion, Cuban presidents and officials used José Martí's words in turn, claiming to re-instil the moral values enshrined in Martí's writings. In the Government of a Hundred Days of 1933,[8] for instance, Antonio Guiteras, President Grau San Martín's Minister of the Interior, launched the slogan '*verguenza contra dinero*' ('shame against money'; Wolf 1969: 264–8).

While Cuban presidents from the early 20th century had promised to re-distribute land to the landless (with the same slogan that Fidel Castro was to use later: 'land to those who work it'), according to Adelfo Martín Barrios, 'The announcement of lands granted by the state was never taken seriously, for the *campesinos*, with their proverbial jocosity, criticized [these] plans, calling them "redistributed lands in *cartuchos* [paper cones]"' (Martín Barrios 1984: 71; my translation). Since the dependent bourgeoisie were in control of neither the state nor the economy, their policies ultimately rested upon the unequal distribution of public benefits (a Spanish *caudillo* tradition) rather than any positive increases in manufacturing or agricultural production (Kapcia 1986: 34).

Unlike the US-backed struggle against economic backwardness and political corruption, the myth of Martí upheld by 'popular' and 'revolutionary nationalists' of the late 19th and 20th centuries was tied to the *mambisa* Fight for social justice and political economic sovereignty. Like many peasant wars (Wolf 1969; Guha 1983), and similar to what was to happen in the Sierra Maestra over a half-century later, the first Cuban revolution was a combined effort by intellectuals and military men, as well as peasants, workers and social protesters of the rural population. According to the rules established by the most 'civilized' of the *independentistas* (i.e. the intellectuals and military leaders), only men in the top ranks of the Cuban revolutionary army with 'gentlemanly standards of masculine honor [*sic*] and civility' had the right to plunder plantations and other property for goods that would later be redistributed (Guerra 2005: 54–5). Mirroring later revolutionary leaders, a small elite group of men had the authority to decide how the needs of the masses were defined and distributed.

But the two main classes in the revolutionary army were united in their reverence for José Martí. Their adoption of Martí's ideas differed from the stultified *martianismo* extracted by the pro-imperialist nationalists. While the latter ignored Martí's fear of US imperialism, the revolutionary and popular nationalists made this into the most important aspect of his work. Like Simón Bolívar[9] before him, who had argued that the Antillean colonies were 'uniquely vulnerable' to US dominance (Abel and Torrents 1986: 12), José Martí was wary of the rise of US influence in the hemisphere, calling for a 'second independence' (Martí 1975: 340) from the United States after Cuba had finally eliminated the authority of the Spanish Crown from its shores. In 1891, Martí warned:

> Whoever says economic union says political union. The nation that buys, commands. The nation that sells, serves. … The nation eager to die sells to a single nation, and the one eager to save itself sells to more than one. A country's excessive influence over the commerce of another becomes political influence. (Martí 1975: 44)

The 'Apostle' came to the heart of the matter when in 1889 he criticized US Secretary of State, James G. Blaine, as a man who was 'purchasable, who true to his character buys and sells in a market of men … under the pretext of treaties of commerce and peace' (Martí 1975: 40–4). During the same period, Indiana Senator Albert Jeremiah Beveridge made a contrasting claim that resembles later US trade initiatives mentioned at the beginning of this chapter: '[American factories] are making more than the American people can use … Fate has written our policy … the trade of the world must and can be ours. … And we shall get it, as our Mother England has told us how' (cited by Thomas 1971: 311–12).

By the late 19th century, the United States had indeed come to dominate the Cuban sugar industry as well as Cuban consumption. The Sugar Trust, for example, was the only supplier of refined sugar to the United States and the only buyer of Cuban crude sugar during its period of existence (between 1887 and 1899; Goizueta-Mimó 1972: 66–7). Just after the Trust came to an end, the first of many US–Cuba trading pacts – the McKinley Tariff Act of 1890 – was ratified, under which the United States established bilateral relations with Cuba, eliminating all tariffs on sugar imports as long as the island agreed to import a growing number of commodities manufactured in the United States (Pérez 1998: 29). The pact was annulled in 1894 with the Wilson Tariff, but bilateral trade agreements like the McKinley Act would be passed between the two countries throughout the first half of the 20th century.

When Spain and the United States signed the Treaty of Paris on 10 December 1898, the United States added political control – in the form of over 10 years of military occupations to control public unrest after the election of US-backed presidents[10] – to its economic and, increasingly, cultural control over the island. Not a single Cuban representative was invited to this meeting between two powerful states, an exclusion that would be repeated over 60 years later when the United States and the Soviet Union settled the Cuban Missile Crisis in October 1962. The US occupation of Cuba, which was justified by the 1898 annihilation of the US battleship *Maine* (blamed on the Spanish), established Cubans' 'shared sense of a stolen destiny' (Kapcia 2000: 87). Cubans' idea of historical injustice was settled when the United States mandated that the Platt Amendment be added to the first Cuban Constitution of 1901, ignoring large-scale protests and anti-US demonstrations across the island (Pérez 1995 [1988]: 187). Based on the moral assumption that Cubans were 'a racially heterogeneous bunch of illiterates unfit to govern themselves' (Aguilar 1998 [1993]: 39), the amendment stated that Cuba

> would make no treaties impairing her sovereignty; contract no foreign debt without guarantees that the interest could be served from ordinary revenues; granted the US the right to intervene in order to protect Cuban sovereignty and a government capable of protecting life, liberty and property; and allowed the US to buy or lease land for coaling or naval stations. (Wolf 1969: 253)

The Platt Amendment defined the political terms of Cuba–US relations until it was eradicated in 1934 by a new generation of reformist (and underground radical) leaders who started the slogan 'Cuba for Cubans'. Despite this radical tradition of Cuban nationalism, economic and cultural ties to the United States continued to strengthen in the 20th century via bilateral trade agreements. The first of these passed during the republican period was the Treaty of Commercial Reciprocity of 1903, which was based on similar exchange relations instituted by the McKinley Tariff Act of 1890. While US politicians and businessmen justified these trade agreements with moral arguments like Cuba's 'backwardness' and need for economic 'progress', revolutionary nationalists like Manuel Sanguily (a former colonel of the revolutionary army and representative of the Constitutional Convention) questioned the moral value of so-called 'reciprocal' trade agreements with the United States. At a meeting to discuss the Treaty of Commercial Reciprocity, Sanguily insisted that

> from the point of view of reciprocity, the name of this [treaty] is inappropriately applied, because we receive from the US a tariff reduction of 20%, and

they receive, in exchange, from us, a progressive series of benefits; given that the totality of that which they give us is surprisingly inferior to that which we give, it is incorrect to consider this as a treaty of true and equitable reciprocity. (Cited by Depestre Catony 1987: 15; my translation)

Like Martí who preceded him, Sanguily protested against US claims to Cuban sugar in terms of a more profound critique of pure economic reason, the kind of economic individualism that disregarded all other social values embedded in a word like 'reciprocity'. By contrast, US military governor Leonardo Wood saw the 1903 Treaty of Commercial Reciprocity entirely in terms of the market value the island would reap 'when' it became the property of the United States:

With the control that we have over Cuba, a control that, without a doubt, will soon be converted into a possession, shortly we will control practically the entire global sugar market. I believe that this is a very desirable acquisition for the US. The island will become gradually Americanized, and in a short time we will have one of the richest and most desirable possessions in the world. (Cited by Depestre Catony 1987: 15; my translation)

Like Wood, pro-imperialist nationalists in Cuba (such as the first president of the republic, Tomás Estrada Palma) assumed the United States would eventually annex Cuba, integrating the island into its 'modern' economy. By contrast, Sanguily and other popular or revolutionary nationalists argued for a 'moral republic' over an 'imperial republic' (Ibarra 1986: 88). Following Martí, counter-hegemonic political groups saw the Cuban peoples' dependence on US commodities as a threat to the very ideas and values that defined Cuba as a nation of revolutionaries. As Martí had written over a decade earlier:

[A]n excessive love for the North is the unwise but easily explained expression of such a lively and vehement desire for [material] progress that [Cubans] are blind to the fact that ideas, like trees, must come from deep roots and compatible soil in order to develop a firm footing and prosper, and that a newborn babe is not given the wisdom and maturity of age merely because one glues on its smooth face a mustache and a pair of side-burns. Monsters are created that way, not nations. (Martí 1975: 52)

Ongoing debates over Martí's ideas involved the contrast between materiality and dependence, on the one hand, and a kind of Cuban spirituality, on the other. These two sides, in turn, rested on a wider moral debate about the very nature of what constituted modern political economy and society. Counter-hegemonic movements, such as Fernando Ortiz's Junta Cubana de Renovación Nacional, emphasized the bipolarity in Cuba between a united and moral notion of what Ortiz called *cubanidad* (Cubanness; Ortiz 1993 [n.d.]), and a greedy, price-wary multitude tied to the sugar industry and dependent on the United States. And, as we have seen, this bipolarity was understood in internal as well as external terms, for many Cubans saw *themselves* as more guilty of causing the '*dolor patrio*' (national pain; Depestre Catony 1987: 33) than people from 'the North'. In the words of former slave Esteban Montejo: 'The Americans wheedled their way into possession of Cuba, but they don't really deserve all the blame. It was the Cubans who obeyed them who were the really guilty people' (Montejo and Barnet 1993 [1968]: 240).

In his monumental work, *Cuban Counterpoint: Tobacco and Sugar* (1995 [1940]), revolutionary intellectual and anthropologist Fernando Ortiz encapsulated values embedded in Cuba's creation myth, particularly humility and agrarianism. In his book, the historical dualism between morality and money is metaphorized in terms of traditional and modern agricultural sectors in Cuba, respectively. Ortiz's emotional argument is that capitalism, epitomized by the growing impersonality of the sugar industry, is not 'Cuban by birth' (Ortiz 1995 [1940]: 17, 36–9). Two commodities – tobacco and sugar – become metaphors for a more profound contrast between Cuba as a place where land and its products are valued both economically *and* socially, and Cuba as an impersonal space driven by prices and markets. Like his intellectual contemporaries[11] and later political leaders such as Ernesto Che Guevara, Ortiz romanticizes Cuban peasant culture. Most importantly for the purposes of this chapter, he portrays the *veguero* (tobacco farmer) as '*entitled*' to 'his smokes' (or other products of the land), while the 'exploited' sugar worker 'has to *buy* [sugar] in the store' (Ortiz 1995 [1940]: 55–6, 84; my emphasis). The difference between entitlements and sociality, on the one hand, and the impersonality and indifference of the market, on the other, is mirrored in one lengthy passage, which relates the way tobacco and sugar are measured:

> [The] … lack of measuring in the case of tobacco leads to a curious consequence in mercantile ethics. As leaf tobacco has no stipulated measure, the shrewdness of the growers and buyers who deal in it must … likewise [be] immeasurable. In these transactions in leaf tobacco there is no market price fixed by a foreign stock exchange, nor uniformity of varieties, nor of chemical composition, nor of volume or weight. For this reason the buying and selling of tobacco requires highly expert personal negotiations between men who understand the complexity of the merchandise they are dealing with, in addition to the general market conditions. (Ortiz 1995 [1940]: 36–7)

While Ortiz admits that the Cuban export sector must utilize concepts and institutions of the market, he also argues that smallholders in Cuba are morally superior to the dispassionate managers or owners of sugar mills and factories. In the lines that follow the above citation, Ortiz points to the corrupt business of the latter as opposed to the moral and social workings of tobacco traders, who must earn the trust of poor *vegueros* (a similar moral argument was once made about 'just' prices in the capitalist economy, to be set by 'the Englishman's word'; Hart 1986: 645; Dilley 1992):

> There is no type of trading that affords such a margin for subtle fraud as that in tobacco. For this reason the good repute of a dealer, based on past dealings, is almost a *sine qua non* for staying in business and doing well at it. ... [T]he sugar industry, on the other hand, is laced with fraud at every step, from the scales on which the cane is first weighed to the shrinkage in weight and the [separate part] that the buyer falsely claims from the warehouse at the port of embarkation, to the refineries who bleach and market the final product abroad. ... [A] deal in sugar had to be written out on paper, with tobacco a man's word was enough. (Ortiz 1995 [1940]: 38–9)

From the force and symmetry of such polemical arguments came the myth of a revolutionary national identity. Thus as we have seen, Cuba's enduring creation myth is based on a moral contrast between a revolutionary or Cuban 'conscience' open to local production and diversification, sociality, hard work and the home-grown 'entitlements' of each Cuban citizen, and the 'existence' of a large and often corrupt group of dependent bourgeoisie (Martí 1959). Such conflicts between national political economic groupings in 20th-century Cuba reflect 'deeply political and moral questions related to the nature of the national community, power and authority, meaning and identity, inclusion and exclusion and distribution of benefits. In short, what was just?' (Fernández 2000: 43).

Town and country in the early republic: pre-revolutionary values of commodities and culture, work and reciprocity

In order to address fully the moral dilemma of Cuban nationhood during the republic, which separated a national 'conscience' from a purely economic 'existence' (Martí 1959), and to understand how post-1959 Cuban leaders incorporated such values into their political and moral economic agendas, we must return to the issue of Cuba's origins as a society divided by city and country. The distinction coincides with the idea that 'culture' (or civilization) is located in the modern sector, which finds its centre in both the polis (Havana) and the metropolis (Spain or the United States). The idea of civilization, as linked to ideal standards for behaviour (Elias 1969, 1982) – or what I will henceforth refer to as 'culture' (as do Cubans themselves) – found material form in imported luxuries, at least prior to 1959. Like other values such as those of work and reciprocity, the criteria for 'cultured' people altered after 1959, and in later chapters we shall see how these moral changes have entered into everyday life.

Similar to the 'high culture' of San Juan (the capital of Puerto Rico; Williams 1990 [1973]), in pre-1959 Cuba the national aesthetic of people in Havana and surrounding areas was linked to an 'international upper-class culture' (Steward 1956: 9–10), epitomized by the consumption of commodities from the United States. Indeed, at least from 1870, when Spain's protectionist Navigation Acts were abrogated and free trade was permitted on the island, plantation society, and later, 'factory central' (Mintz 1956) society was a consumer society, increasingly tied to commodities from the North. The values that emerged from such relations were to become a fly in the ointment for post-1959 intellectuals, whose plans opposed those upheld by the United States and pro-imperialist Cuban elites.

Parallel with the urban poor, for lower classes in rural areas of the Havana Province (like Tuta), 'the ability to use behavior [sic] patterns and symbols of the prominent people' (Scheele 1956: 426) developed with increased access to and knowledge about commodities transported from Havana. Most of these commodities came from outside Cuba, as did the 'patterns and symbols of … prominent people' (Scheele 1956: 426) living in foreign places. Indeed, for middle and upper classes in Havana, education was 'finished' by attending universities in the metropolis (Knight 1990: 81). As in other former colonies (particularly in the Caribbean), the values inculcated in urban areas became the culture of the modern citizen.

For *habaneros* especially, the United States's presence affected nearly all aspects of material life, from the price of sugar to a general preference for pricier bread made from imported wheat. Indeed, as Anne McClintock (1995), Timothy Burke (1996), Richard Wilk (2006) and other post-colonial scholars[12] have argued for other contexts, one key feature of colonialism and related power structures has been the link between the 'civilizing mission', and the 'cultivation of demand for imported consumer goods' (Wilk 2006: 17). In Tuta, this demand was evident when people spoke of 'good' commodities available in Havana or from intermediaries. As one elderly Tutaño recalled, drawing a parallel between past and present relations between polis and hinterland:

> It has always been that if one wants to look for something *good*, they have to go to Havana. … We *campesinos* [peasants] have always suffered because people who come from the city with quality things sell them at exorbitant prices. But people buy them anyway.

Though only one in seven Cubans lived in Havana prior to 1959, its power over the countryside was so great that Ernesto Che Guevara famously compared Cuba to 'a dwarf with enormous head and swollen chest' whose 'weak legs or short arms do not match the rest of his anatomy' (cited by Wolf 1969: 261).

US involvement in Cuba exacerbated pre-existing bipolarities between Havana and rural Cuba, particularly as a result of US ownership of lands previously occupied by Cubans. Agricultural workers, who soon comprised the majority of working-age rural Cubans, grew more and more dependent on the sugar industry for their survival. They suffered most during the post-harvest 'dead time' (*tiempo muerto*), the season when work ceased on the cane fields. By contrast, political elites in Havana signed 'reciprocal' contracts with the United States, often giving in to pressure to accept lower prices for sugar when the United States threatened to cease supplies. Such policies threatened the livelihoods of the rural proletariat and the *colonos*, or sharecrop farmers who produced sugar cane for national and foreign mills (Le Riverend 1966: 159). Though many in the city encouraged US investment and trade, developing 'new thirsts which could be slaked only with US goods' (Thomas 1971: 600), a large population in the countryside, especially in the east, was left landless and dependent on the sugar industry for both commodities (usually bought via credit at stores located in sugar complexes) and limited periods of work during the sugar harvest.

An early sign of misery in the countryside was the 1912 uprising of Afro-Cubans in the east. Since the destruction of the War of Independence, Afro-Cubans had migrated to the more rugged and less desirable terrain of the eastern Sierra Maestra to find autonomy in farming or petty commodity production. In 1912, over 10 000 blacks set fire to sugar lands, local stores run by usurious merchants and public records that legalized post-republic property relations (Pérez 1986: 533). According to Pérez, this *lucha* (as it was called) represented the first collective response to US encroachment on Cuban land (Pérez 1986: 538–9). Despite the unified sentiment of injustice, the aftermath of this protest led to heightened racial tensions between rural whites and blacks; indeed, many of the former adopted racial prejudices that Martí had condemned in the name of national unity (Martí 1975, 1977). In a brutal retaliation after the uprising, the Rural Guard cut off the heads of Afro-Cubans, leaving them in public places (Pérez 1986: 537), ultimately killing over 3000, many of whom were not even implicated in the insurgency (Knight 1990: 238–9). Collective memories of 1912 and similar injustices were to motivate Afro-Cuban people in the east to join Fidel Castro's Rebel Army in the late 1950s. More than radical Marxism (Al Roy 1967), they were convinced by the rebel army's policy to create a 'free territory' in the Sierra Maestra region, re-distributing lands as well as food and other necessities to people who had long lived a precarious existence (Yaffe 2009: 17).

Many of my Afro-Cuban informants, as well as the former slave Esteban Montejo, saw the rise of US dominance over the island as a serious blow to the Afro-Cuban population, who had formerly been recognized as humble revolutionaries, if not entirely equal to the gentlemanly heroes of the War of Independence. In Montejo's words:

> I know that 95% of the blacks fought in the war, but they [the North Americans] started saying it was only 75%. Well, no one got up and told them they were lying, and the result was the Negroes [who were left without military pensions] found themselves out in the streets – men as brave as lions, out in the streets. It was unjust, but that's what happened. (Montejo and Barnet 1993 [1968]: 236)

Afro-Cubans found ways of asserting their self-worth, however, and this was often by turning away from agriculture to find jobs in the city. Moving closer to the polis increased the desire to consume commodities imported from the United States. Indeed, as we have seen, dominant consumers living in Havana, such as expatriate North Americans, largely set the standards for subordinate people who aspired to 'the good life'.

> Havana was both the point of entry of American influence and the chief link between the island and the society and economy of the American continent. Showing great contrasts between its middle and upper classes, geared to American-style ideals of mobility and consumption, and the urban poor, it nevertheless demonstrated in its ambience and lifestyles the magnetism of the American 'way of life'. (Wolf 1969: 261–2)

Like generations before them, many elderly people in Tuta made a direct correlation between increased trade with the United States and 'the good life'. Referring to recent trade agreements between Cuba and China, one man in his eighties said: 'Every time that trade with the US increases, Cubans live better. ... The people have more trust in US products. ... People eat food that comes from the North with more confidence than they do eating the *same* food that comes from China'. In another interview, a woman in her seventies told a joke that had been popular in Tuta prior to the 'triumph' of the 1959 Revolution. If not rib-tickling, it is certainly illuminating in this context:

> There was a woman who was the talk of [the] town because she had spent her whole month's savings on a facial cream that came from the United States. She applied the cream first thing in the morning and last thing at night. One day I saw her and her face was black, though she was usually coffee-colored. It was morning and she had just arisen to perform her daily ritual of applying the skin cream. I made her show it to me, and she demonstrated how to apply it, but was very sparing because, as she emphasized, it came from the US and was very expensive. She said to me, 'you have to rub it on your nose like this'. It was then that I realized it was not cream at all, but pig shit!

Typical of Cuban irony, analysed in more detail in chapter 3, this joke plays with a given set of values, which are then overturned at the end. The most interesting aspect of this joke and others like it is that Cubans themselves realize the contradictions of over-valuing US commodities. Indeed, the joke highlights a general problematic of consumption in most post-colonial societies, especially pronounced in the Caribbean region because of its proximity to what Pierre Bourdieu (2005: 225) calls the most 'globalized' economy in the world. As with other people in the 'developing' world, so many Cubans care about the origins of products, and those made in the United States are often considered the best. My Cuban 'mother', for instance, preferred expensive, packaged hot dogs imported from the United States over homemade chorizo. Values such as these affect how the ethnographer is served as a guest in the home. Because the foreigner is linked to the outside, he or she is associated with imported commodities. Thus while anthropologist Richard Wilk's mouth watered at the prospects of eating 'real Belizean food' (Wilk 1999; 2006: 199) such as coconut dumplings and fish, as an honoured guest he was served 'the best food the house could afford: canned corned beef, packaged white bread and a luke-warm 7-Up'. Likewise, at the beginning of the fieldwork period, my host family attempted to appease my 'superior' western tastes by spending a week's wages on canned fruit cocktail, soda and hot dogs, all sold in hard currency.

The dual economy that precludes many Cubans from frequenting hard currency markets has not always been so exclusive. A man who formerly worked for the nearby sugar complex, for instance, spoke of a time 'Before [when] … the American factories had tremendous snacks! And they were cheap too! One peso could buy a mountain of things', opposing this to now, when 'those who have the dollar live the best'. Elderly Tutaños recalled the importance of US foodstuffs and other commodities in their lives 'before' (the Revolution). The 'before/after idiom' (Holbraad 2000: 1–3) has been used more recently to describe times preceding and superseding the 1990s' economic crisis, and I return to some of the ways it is used in chapter 4. Here it is important to consider the words of Tutaños who remember commodities imported from the United States during the pre-revolutionary period, if only because these memories are still very important for how Tutaños value things. As anthropologist David Sutton argues in his work on food and memory on the island of Kalymnos, Greece, 'if "we are what we eat", then "we are what we *ate*" as well' (Sutton 2001: 7; my emphasis). Sutton's writings reveal that the form in which memories are cast, such as eating, is often more important than their content, and that subaltern histories may be more readily interpreted if we pay attention to how memories relate to embodiment (Sutton 2001: 10–12).

Such insights are relevant to my later analysis of how Tutaños value food commodities, but are also helpful when considering the way Tutaños remember another form of cultural performance: work. In fact, values of the two – commodities and work – are often opposed in Cuban historiography. For instance, Martí contrasted the wealth that US commodities represented with a kind of spiritual wealth earned only through hard work. In so doing, he counteracted the Spanish (and Veblenian)[13] tradition upheld by the 'cultured' Cuban elite, who associated manual work with poverty. Instead, Martí regarded manual labour as *more* valuable than material or monetary wealth, as it alone allowed 'man' to develop into a truly spiritual being:

> Man grows with the work that comes from his hands. It is easy to see how the idle are impoverished and degraded in a few generations, until they are but fragile vessels of clay, with feeble legs which they cover with sweet-smelling perfumes and patent boots. ... Some parents instill in the child a passion for wealth, oblivious to the fact that the only durable wealth results from the sweat of the brow. (Cited by Ibarra 1986: 85, 89)

Though Martí's idea of work was to become an essential aspect of the national moral economy in Cuba (particularly through the writings of Che Guevara, addressed below), we have seen that it was not relevant to all Cubans. Indeed, in late 19th- and 20th-century Cuba, a large section of the population adopted the 'cultured' idea of work, according to which manual labour was lowest on the hierarchy. The value of non-agricultural work in urban areas was especially important for those who had formerly worked the land as slaves. As Franklin Knight has written regarding the period of slavery in the Caribbean, so for the subsequent period in Cuba:

> The rural exodus of the free [blacks] might ... have had powerful subconscious motivations. The social mores of Caribbean plantation society consistently denigrated those who did manual and menial labour. For this reason, the plantation field slaves represented the lowest social category. And in an atmosphere where the correlation between black skin and the status of slavery was almost taken for granted, free nonwhites in rural areas faced the unavoidable and continuous problem of proving their freedom. Increasing the social distance between slavery and freedom for the Afro-Caribbean population might have meant increasing the physical distance as well. (Knight 1990: 141)

Like other rural inhabitants who associate 'the good life' with the polis and not with the fields (Redfield 1956: 133), so many former Cuban slaves found that their quality of life was much better in the city. There they could establish secret societies (*negros curros*) and centres to practise African religions like Ifá, as well as social clubs where Afro-Cuban dancing and music continued to thrive.

One of my elderly friends in Tuta concurred with the idea that Afro-Cubans of the Cuban republic had no place 'in the fields'. When I asked him whether there were many blacks in Tuta who worked in agriculture, he responded: 'Blacks in Cuba do not like working in agriculture. In Tuta we did everything we could do to move to Havana. There we could find better positions –

tailoring, laundering, building ... Agriculture is not for the blacks'. Despite limited opportunities in the city, blacks who moved to Havana still faced limitations in how highly 'cultured' they could become. While some (usually 'mulattoes') did enter into political and social ranks, the colour of their skin was an obstacle to social mobility. Another of my elderly friends in Tuta highlighted the challenges of being an Afro-Cuban in the republic when he told me the story of how he came to be a mechanical engineer in the 1940s:

> I was very privileged to go to the best school for engineering in the Havana Province, which is actually the best school in the whole of Latin America. ... It had international students from Jamaica and other places. Anyone who left this school had the door opened for them. But when I finished my training and went to get a job with the national job agency in Havana, they would not give me work [He rubbed his arm]. *La raza* [the race]. They told me, 'No. There is no job right now, come back later', and left me there. There was a job going for a mechanical engineer. They eventually gave it to a white man who had never even received his credentials.

Afro-Cubans of the republic could never rise to top positions, for whiteness was a pre-requisite for associating with the most 'cultured' class in Havana. For their part, upper classes in Havana associated culture with US commodities but also with leisure. Thus upper class *habaneros* were required to 'relax sufficiently from money-grabbing to enjoy spiritual things, to cultivate such social graces as Latin American singing and dancing' (Scheele 1956: 425). Moreover, as Raymond L. Scheele notes for San Juan, so for Havana, Spanish-born elites saw 'work' in both economic *and* social terms, contrary to the North Americans, whom they regarded as 'uncivilized materialists' (Scheele 1956: 425). Such a view of work was partly maintained in the less 'cultured' countryside, where both social relations and household economies were maintained through hard work. Thus according to another elderly Tutaño, 'before' the Revolution work was valued insofar as it maintained the household as well as one's dignity:

> A good worker always had enough. I always got food from people for whom I worked. ... The worthless [*aquellos que no sirvían*] did not work. If you worked for rich people and today take a biscuit and tomorrow eat a steak and the next day come back with their clean clothes, at the first opportunity they will fire you. ... If I need work, I have to respect your things. Because if not, in the end, they know. ... I worked for the richest people in the world. ... I looked at the things they had but didn't take anything. ... I cannot have your things because you are rich and I am poor. But you came to my house and it was clean and had the same scent as yours.

The words above not only indicate Tutaños valuation of hard work, but also the idea that, through work, individuals may obtain all the commodities regarded as necessary for the household (such as quality cleaning products that smell nice), in the process, securing non-material values such as self-respect and autonomy. The power of such memories is their contradistinction to 'now': the period of scarcities that began in the early 1960s and escalated in the post-1990 period after the fall of the Soviet Union. As mentioned above, many elderly Tutaños spoke of the time 'before' (1959), when one could procure basic household needs (*necesidades*; Holbraad 2000), such as foodstuffs, through hard work. Consider the following quotes extracted from different interviews with older Tutaños:

> Before in Tuta there was a large number of consumer goods – foods, clothes … Now one has to pay 38 pesos for *one* pork steak! Before wages were enough to pay for food. Now, no matter how much one works, they cannot eat!

> Before my dad worked for the American textile factory. We never had a problem getting anything to eat … After, when they started redistribution, there was much less food. It was the *end* of beef, the *end* of pork. … They sent food to hotels, hospitals, schools … They started marking cows' ears, even those of private owners. … And milk from the cows was reserved only for children … Now there are small portions – you only get a tiny piece of fish in the *bodega* [local distribution centre] for three people! We only get five eggs a month. The *mortadella* [minced pork] is of a very bad quality.

> Before I had more than enough money. I had enough clothes, enough shoes. Now, if I want food, I have to go to someone who comes home from work and asks, 'do you want eggs? Do you want minced meat?' … I have to buy *por otra vía* [another way; illegally]. In the *carnicería* [meat market] they give you a *gota* [drop], nothing more. … Before people say that we lived poorly. But they lived with their salary. With my *father's* money, with my *father's* work, he built this [his daughter's] house. *This* house had to be finished with the help of a foreigner. Now the people must live through gifts.

These quotes are interesting because they reveal Tutaño ideas of reciprocity – by working hard, one may acquire 'enough' food for the household (Holbraad 2000: 4). After 1959, such ideas of reciprocity 'jumped scale' from the household to the nation and, in chapter 4, I address this nationalized version of reciprocity and its relation to more recent ideas of reciprocity and redistribution in Tuta. The citations above also highlight the problem of money in post-1959 Cuba. Continuous with Martí's opposition between the value of commodities and that of hard work, and in line with the socialist contract between citizen-workers and the state, Che Guevara's model for the economy questioned the value of money and commodities as measures of national progress. As Cuban economist and biographer of Guevara, Carlos Tablada, argues (1990 [1987]: 24), Che initiated a shift in the Cuban morality of work, from personal or familial interests to the duty to work 'for the benefit of society'. This jump in scale was justified not only in terms of a broader, nationalized idea of work and reciprocity, but also in relation to an alternative view of culture, which I address in chapter 5.

Agrarianism, Guevara and the Great Debate

In the decade after the 'triumph' of Fidel Castro's Revolution, Marxist values of work and social property became increasingly associated with nationalist values of agrarianism, self-sacrifice and humility. As Fidel said of his 26th of July movement, which was to overthrow the dictator Fulgencio Batista, '[This is] the revolutionary organization of the humble, for the humble and by the humble…' (Thomas 1971: 874). In an auspicious show of solidarity with the *humildes* of Cuba's creation myth, Fidel had his family's lands in the east burnt to the ground as 'an example of economic warfare against the rich of Cuba' (Szulc 1989 [1986]: 10). And one of the first policies he implemented after coming to power was to abolish discrimination in public places, such as white-only beaches and clubs (Silverman 1971: 7; Knight 1990: 245–8).

Leader of the 26th of July movement, so named after the failed insurgency on Moncada Army Barracks on 26 July 1953, Fidel Castro correlated his actions with the revolutionary *Lucha* unified by José Martí (Figure 2.2). For instance, he identified his army as the 'Generación del Centinario' or the generation of the 100th anniversary of Martí's birth in 1953 (Kapcia 2000: 93). Along with early re-distributions in the Sierra, such associations between the Revolution and Cuba's historical Fight brought Fidel Castro and his group mass popularity. The future prime minister and president of Cuba was indeed triumphant when he entered Havana, hand-in-hand with Ernesto Guevara, Camilo Cienfuegos[14] and his brother Raul on 1 January 1959.

> How could Fidel not be convinced that victory was his, and his alone, amid the cheers of hundreds of thousands of Cubans as they hung, fascinated by his oratorical genius, upon his every word, expression, and gesture? … It was only natural that there should be an imaginary reconstruction of events: 'the Sierra' had won, not 'the plains' [urban areas in the east]; the 26th of July had done it all, without any important allies; its leaders had carried the day, thanks to their exceptional wisdom and intuition … the many-faceted, complex process of the Sierra would be transformed into a magnificent, straightforward epic that could be repeated anywhere with bravery and integrity. (Casteñeda 1998: 140)

The popularity of the Castro group was partly achieved through a mythical representation of Fidel Castro's rebel army as the 'new *mambises*' (Kapcia 2000: 96). Like the late 19th-century *mambises*, Castro's army had marched west from the Sierra Maestra; unlike the heroes of the War of Independence, however, the Castro group made it to Havana. Thus the

Figure 2.2   A bust of Martí in front of an agricultural cooperative. Source: author.

triumph of the Revolution was portrayed as the successful 'culmination of Cuba's history'. Likewise, the Argentinian, Che Guevara, was compared to the Dominican Máximo Gómez: both were non-Cubans who became leaders in the Cuban cause for national sovereignty (Kapcia 2000: 171, 183).

Che was for post-1959 Cuban society what José Martí had been for the post-1895 generation (Kapcia 1997: 25–6). Along with Fidel Castro, Guevara re-invigorated Bolívar's transnational quest for regional autonomy from Spain and, then, the United States. Indeed, Guevara saw Cuba as the 'key' that would unlock a hemispheric struggle against the United States (Guevara 1971: 352). According to this logic, it was up to Cubans to initiate the Fight against what Martí had called the 'monster of the North'. Like Martí before him, Guevara was seen as the primary intellectual who was to lead the 'masses' in the new Fight against US imperialism. In fact, *Time* magazine devoted a cover story to Che in the mid-1960s, referring to him as the 'brain of the Revolution', to Fidel as the 'heart' and to Raul as the 'fist' of the Revolution (Casteñeda 1998: 179). Guevara combined his version of Marxism, which emphasized anti-imperialism, with Martí's revolutionary morality: 'We find in Che Guevara an unequalled integration of Martí's insights and Marx's theory, into a revolutionary practice by which people might liberate themselves by becoming aware of their social potentials' (Ken Cole, cited by Yaffe 2009: 69). Central to Guevara's thought was his adoption of Regis Debray's[15] 'guerrilla thesis', which attributed to revolutionary socialism the power of peasants rather than workers. Such convictions were to re-structure long-held 'codes' of Cuban revolutionary nationalism, particularly agrarianism, self-sacrifice, humility and the *Lucha* (Kapcia 2000, 2008).

Like romantic portrayals of the War of Independence, Guevara's guerrilla thesis ignored middle class support for the revolutionary cause, over-emphasizing the role of the 'humble' peasant (Casteñeda 1998: 192). Of the rural Cuban population in 1958, however, just 220 000 were 'peasants' – mostly sharecropping *colonos* – while 600 000 were agricultural workers (Pérez 1995 [1988]: 317–18). The poorest of the minority of peasants lived in the Sierra Maestra:

> In the zones where Guevara came to know the peasants' way of life, land was essential. As a result, agrarian reform was crucial to the process of merging the peasants with the guerrilla camp. That is why Che called the Rebel Army a 'peasant army' and the 26th of July a 'peasant movement'. (Casteñeda 1998: 123)

In the early 1960s, the first two agrarian reforms were partly to reverse this idealization of peasants, particularly the 2nd Agrarian Reform Law, which was more Marxist-Leninist than the first (the latter was similar to other populist agrarian reforms in Latin America at the time). The real and imagined rural origins of the long historical *Lucha* for social justice and sovereignty were again of ideological use in the 1990s and 2000s, however, when the Cuban government instituted a plan for national food sovereignty (see chapter 6). Indeed, as Antoni Kapcia (2000: 236–7) argues, along with *Lucha*, agrarianism is 'possibly the best surviving code in the 2000s'.

The 1990s and 2000s campaign for national food production (and the earlier National Food Programme of 1987–1994) were not the first food security initiatives in socialist Cuba, however. In the early 1960s, Guevara instituted a plan for agricultural diversification. At this time, he was head of the National Institute of Agrarian Reform, a model economic and ideological institution for all sectors of the planned economy (see Valdés 2003). But Guevara's policy for national food security failed within 4 years, initiating food scarcities and the need for rations in 1962. His diversification plan was unsuccessful because of his 'guerrilla habit of favoring [*sic*] exaggerated efforts over agronomic practicality' (Díaz-Brisquets and Pérez-Lopez 2000: 94). Indeed, unlike the 're-peasantization' of the 1990s and 2000s, Che's 1960s farms were too large and specialized, with too many crops (Dumont 1973 [1970]: 114).

Che's mission to 'conquer nature with science and technology' (Díaz-Brisquets and Pérez-Lopez 2000: 10), resembled the Ten Million Ton of sugar policy of the late 1960s (see below): both were based on the large-scale re-structuring of an entire sector. Accordingly, such efforts relied on an energy-intensive model for agriculture. For instance, from 1965 to 1985, imports of agrochemicals from the Soviets increased four times over (many of which were banned in the United States, such as dichlorodiphenyltrichloroethane, or DDT). Moreover, in 1989 Cuba had more tractors than any other country in Latin America: 79 000 (Díaz-Brisquets and Pérez-Lopez 2000: 101).

Similar to western modernists of the 1960s like Walt Whitman Rostow, for Castro and Guevara an increase in technological 'sophistication' and industrialization was the only way to ensure a cornucopian future. Indeed, the model adopted in Cuba was (and is) as concerned with progressive development as the liberal project of modernist development introduced during the same time period (Foster-Carter 1976: 177). In a commemorative speech on 26 July 1968, Fidel Castro promised that 'in the future [with an increase in technology], there will be practically unlimited quantities of material goods' (Castro Ruz 1971: 363). Contrary to the right side of the theoretical spectrum, however, Cuba's modernism problematized the western idea of economic progress, and opposed the kind of development proposed by Rostow in his *Stages of Economic Growth: a Non-communist Manifesto* (1960). While in the Rostowian model, 'the [Cuban] Revolution interrupted healthy capitalist growth', in the model adopted in 1960s Cuba, 'it was a precondition to resolving the contradictions obstructing development by ending Cuba's subjugation to the needs of US capitalism' (Yaffe 2009: 6). The former relied on norms such as self-interested (economically) rational individuals; the latter prescribed for *another* future society based on the centralized redistribution of needs.

Like the more influential world systems and dependency theories of Immanuel Wallerstein and Andre Gunder Frank, respectively, Guevara questioned the universal benefits of economic rationality and commodification, turning away from orthodox Marxism (based on a transition from capitalism to communism) for the anti-imperialism of Lenin and the late Marx and Engels (Yaffe 2009: 55). His was a national planned economy (and, ultimately, a world socialist economy)[16] isolated from the kinds of unequal exchanges Cuba had long experienced with the United States: 'How can it be "mutually beneficial" to sell at world market prices the raw materials that cost the underdeveloped countries immeasurable sweat and suffering, and to buy at world market prices the machinery produced in today's big automated factories?' (Guevara, cited by Tablada 1990 [1987]: 159). Instead of world market prices, then, Guevara advocated a controlled pricing system to ensure the basic needs of all Cuban citizens:

> The fact that a series of basic necessities of life are made available at prices below their cost does no harm to the country's economy so long as the prices

of another series of non-essential articles are raised above their cost. Under socialism, a particular price can be set as far from an item's actual value as is considered necessary. *It is the overall proportions that are decisive.* ... [A] whole series of necessities of life must be supplied at low prices, while we can go overboard in the other direction with less-important goods, openly ignoring the law of value in each concrete case. (Guevara, cited by Tablada 1990 [1987]: 146–7, 154; emphasis in original)

Such a policy had economic as well as social benefits, for in the first year of the Revolution, Cuban tariffs and cuts on luxury imports like televisions and cars saved the Cuban government over US $70 million, savings that could be used to subsidize basic necessities (Pérez 1995 [1988]: 319–20).

Along with his position in the National Institute of Agrarian Reform, Guevara was (not surprisingly) head of the Department of Industrialization. He was also president of the National Bank of Cuba, and it was in this role that he was most successful implementing his pricing policies.[17] Che's first agenda in 1959 was to eliminate all ties to international financial institutions like the International Monetary Fund (IMF). In that year, a law was passed that declared banking a public institution; its primary purpose was not lending, but accounting for all items produced in the country. In 1961, Cuba came closer than ever to Che's goal of ensuring national control over finance and economic planning. On 4 August of that year he passed a law that ended thus:

the insecurity and risks which result from the fact that the banknotes currently in circulation are printed by foreign enterprises outside the effective control of the revolutionary government ... [and] to prevent national monetary resources in the possession of the external counter-revolution from being used to conspire against the revolutionary government and the people of Cuba. (Cited by Yaffe 2009: 29–30)

After the law was enacted, Cubans on the island had just one weekend to exchange their US-backed pesos (up to 200 per family) with new money printed with pictures of revolutionary heroes such as Guevara himself. Flights to and from Cuba were cancelled over the long weekend (4–8 August) to prevent the entry of outsiders with interests in hoarding Cuban money. After this initial step, Cubans, enterprises and other institutions (who had separate accounts) were permitted to withdraw up to 10 000 pesos per month, a policy that prevented capital flight and increased national coffers by 72.5 million pesos (then equivalent to the US dollar).

Guevara's interest in re-structuring Cuban finance to benefit national planning stemmed from his deep belief in a model for the economy in which money and pricing were not the primary forms of value. He explained his theoretical and moral position in writing, during what Ernest Mandel called the Great Debate of 1963–1965 (Silverman 1971: 1). The primary question that sparked the debate was: how quickly can a revolutionary society eliminate the law of value, which determines prices and, in turn, supply and demand in a capitalist economy? On one side of the dialogue were the advocates of market socialism (i.e. Carlos Rafael Rodríguez, E.G. Liberman, René Dumont, Charles Bettleheim),[18] who promoted the self-financing of state enterprises, the use of material incentives and the preference for small and medium farmers over large cooperatives or state farms. Their primary argument was that, due to Cuba's position as an 'underdeveloped' nation, it could not instil a communist consciousness in its population until the economic 'base' was fully formed (Meso-Lago 1981: 39–47). In other words, according to the market socialist theorists, 'abundance' should come first, with the help of market mechanisms to determine supply and demand. Only then could the 'inner wealth' of Cuban men and women be developed, along with a centralized communist economy.

But the 'revolutionary economist' (Tablada 1990 [1987]: 91) did not see a contradiction between the promise of future abundance and his emphasis on immaterial values. According to Guevara's call to create 'both socialism and communism at once', each aspect of development would work in unison with the other. This socio-economic teleology was based on the 'formation of the new human being' (Guevara 1971: 347) who works voluntarily and sacrifices his material desires for the greater 'common good'. Apart from hard work, the creation of what he called the 'new man' would come from a *'guerrillerista* ... commitment to constant mobilization, mass involvement, struggle, defense, unity' (Kapcia 2008: 35). This sense of militarization and national defence, heightened by the failed 1961 US invasion of the *Playa Girón* (Bay of Pigs) and internal counter-revolutionary violence, empowered ordinary Cubans to 'be like Che'. Mass work mobilizations to construct houses, schools and hospitals, for instance,

> succeed[ed] in kindling in the population a sense of togetherness and defensive unity, and, rather than creating a coercive militarist atmosphere, seemed to have created a sense of collective struggle that gave ordinary Cubans a sense of empowerment and even quietly 'heroic' participation in the defense of the Revolution. (Kapcia 2008: 152)

Though market socialism was to become the primary model of the Cuban economy in the 1970s, when Cuba was tied to the Soviet Union both economically and politically, in the mid to late 1960s, and arguably from 1986 to the present, Ernesto Guevara's model was the most influential policy for both economic and moral development in Cuba. Following the earlier writings of Marx, which addressed the social liberation of workers (Kapcia 1997: 30), Guevara set out to change both the morality of the Cuban people and the economy at the same time. When, as he argued, the 'conscience' of people was lifted to meet the demands of revolutionary society, their work effort would increase as would overall output (Guevara 1971: 342).

But the emphasis on an economic *and* moral Revolution necessitated a vanguard to guide the 'uncultured' people into spiritual and material prosperity. Continuing the gentlemanly tradition of the first revolutionary nationalists, Guevara assigned to the Party all aspects of social reproduction, and prioritized a combined education–work programme for Cuban youth, whom he saw as untainted by the 'original sin' of capitalism (Guevara 1971: 394):

> In our [new] society, the youth and the Party play a big role. The former is particularly important because it is the malleable clay with which the new man, without any of the previous defects, can be formed. ... Education is increasingly integral, and we do not neglect the incorporation of the students into work from the very beginning. Our scholarship students do physical work during vacation or together with their studies. In some cases, work is a prize, while in others it is an educational tool; it is never a punishment. (Guevara 1971: 342, 351)

Like Martí, Guevara and Castro saw manual labour in positive terms, a 'prize' that all shared, including the revolutionary leaders who became famous for working alongside ordinary people in cane cutting and other arduous labour. Indeed, though he claimed that agricultural labour would not be an issue in the future, when technological innovation would bring Cuba into the modern age (Carciofi 1983: 196), Che encouraged volunteerism and work mobilizations in his drive for the total change of 'man'. In the late 1960s, the policy of work mobilizations and moral incentives was essential for Castro's Ten Million Ton sugar campaign of 1970, when collective work in the cane fields was conceived as a 'national campaign for the construction of socialism' (Carciofi 1983: 199). The shift from Guevara's earlier model of diversification to the mass production of sugar for export (now to the Soviet Union) was Castro's compromise, based on the implementation of a planned economy (until 1970) and justified by Guevara himself in terms of his view of justice. Thus, rather than an unequal exchange of commodities between rich and poor countries, Cuba's barter terms of trade with the USSR was seen as beneficial to both countries and to world socialism as a whole (Yaffe 2009: 172).

Similarly, Guevara saw exchanges of goods between state-run enterprises as non-commodified, their prices only serving the purpose of recording production:

> [I]n exchange between state-owned production units there was no transference of ownership. For accounting purposes only, the 'delivery of products' was accorded a 'price' and relevant adjustments were made in enterprise accounts held in the Treasury. (Yaffe 2009: 54)

With the aim of reducing the importance of money as a means of earning wages or buying commodities, Guevara's model called for an elimination of the capitalist law of value. According to the latter, 'each individual consumer possess[es] the ability to "vote," by income spent, on what gets produced in the market [and on its price]' (Baber 1987: 48). By contrast, Guevara called for an eradication of the price mechanism and, ultimately, an elimination of money altogether. In his quest to create a society of 'new men' of socialism, he advocated a strict break from the capitalist past, attempting to

> eliminate ... the residues of an education and an upbringing systematically oriented toward the isolation of the individual, ... [and] commodity relations. The commodity is the economic cell of capitalist society: as long as it exists, its effects will make themselves felt in the organization of production and therefore in man's consciousness. (Guevara 1971: 341–2)

Even in the wording of his economic policy, 'commerce' was replaced by 'the delivery of products', 'profit' with 'the record of results' (Yaffe 2009: 85). Responding to the market socialists of the Great Debate, Che argued that the law of value was no longer relevant for a revolutionary society such as Cuba, as 'a society in Revolution creates its own rules' (Kapcia 1997: 32).

In a speech on 26 July 1968, Castro followed the 'brain of the Revolution' in contrasting the 'new society' that was in formation to that of pre-1959 Cuba, when 'imperialists ... [could] bu[y] off revolutionary leaders with a few miserable dollars' (Castro Ruz 1971: 371). By doing so he correlated the polarity between these two periods with another opposition between two forms of wealth: that represented by money and desirable commodities and that earned when one became a member of a redistributive society:

> Families ... are beginning to find that the money they cherished before, because it meant the health of their child, the bread of their child, the medicine of their child, the recreation of their child, the education of their child, is losing its meaning. ... Money continues to be used for other things, but for these things it is becoming increasingly unnecessary. To enjoy oneself, to take a trip, to drink a beer, for any of these things, all right; and people value these, but they value more the health of their child, the education of their child, the bread of their child, the roof over the head of their child. In other words, all the essential things, the things they value most and for which they sacrificed recreation, beer and all the rest, are no longer obtained with money. ... Money will have less and less meaning. (Castro Ruz 1971: 376)

As this illustrates, the most influential model of the Cuban economy in the post-1959 period was based on a subordination of individual desires for commodities (e.g. beer) to collective assets owned by the Cuban nation as a whole, such as nationalized healthcare and centralized food distribution.

The underlying values of collective over individual interests in the model called for a re-ordering of the Cuban nation into one big enterprise. Raw materials would be exchanged for manufactured goods between enterprises rather than money and commodities, and material incentives (or wages) would be replaced by moral incentives and rations, though the latter, too, would become unnecessary over time, when work input by the population balanced with centralized output. According to this spatio-temporal scheme, the socialist national economy, which emphasizes access to work, would eventually be replaced by the national (and, then, international) communist economy, based on the tenet 'from each according to his ability, to each according to his needs'. When work itself ceases to be a commodity, as material incentives are eventually replaced by moral incentives and as production enables collective abundance, then there would be little need for money at all.

As Juan Martínez-Alier argued, however, 'though material incentives would produce "two types of men," so would moral incentives – those who decided what will be produced and those whose production efforts and standards of living would be decided from above' (Martínez-Alier 1977: 164). Thus while campaigns for nationalization, healthcare, literacy, rationing, equality for blacks and women and volunteerism led to a drastic shift in privilege from Cuban elites (most of whom left the country) to Party members and then, to the population at large, the change from a society dominated by market values to a centralized system allowed little space for local valuations of goods deemed 'unnecessary'. In later chapters, it will become clear that local people in Cuba developed ways to counteract top-down norms for economic behaviour, just as they had done before the 1959 Revolution.

Pendulum shifts and moral continuities

Guevarian and Martían ideals remain essential for understanding the political and moral economy of present-day Cuba. While Martí shaped Cuba's creation myth prior to 1959, after this date Cuban history has largely coincided with what Helen Yaffe (2009: 262) calls the 'Guevarista pendulum ... between what is desirable and what is necessary – with Guevara's ideas being associated with the vitality of the Revolution' (see appendix 1 for a timeline of key political economic events of the Cuban Revolution from 1959 to 2013).

As we have seen, the first swing towards Guevarian moralism was in the mid-1960s. A swing away occurred in the 1970s, after the Ten Million Ton sugar campaign failed and drained all other productive sectors of manpower, funds and moral impetus. The adoption of the Soviet market socialist model coincided with a historic upsurge in the price of sugar, set at 65 cents per pound in 1974 (Knight 1990: 251). Though reflecting Cuba's continued dependency on global market trends, the high price of sugar had the effect of shortening lines for consumer goods, of which there was a much larger selection than that available in ration books (*libretas*). Ordinary people recalled the period as a time of abundance: 'For one brief shining moment it seemed that the good times had finally arrived, justifying the revolutionary sacrifices of a decade' (Knight 1990: 251). By the early 1980s, however, the price of sugar had fallen to 6 cents, and in 1986 Fidel Castro's 'Rectification' campaign signalled a correction of the 'errors' of the 1970s and a re-adoption of Guevarian ideas of volunteerism, self-sacrifice and moral incentives. As with Mao's Five-Year plans in China, the shift from material back to moral incentives led to a renewed period of scarcities.

After the fall of the Soviet Union, scarcities of food and energy supplies became critical, leading to another pendulum shift away from Guevara's moral economy. In 1991, a period Fidel Castro referred to as the 'Special Period in Time of Peace' began. State supplies plummeted and prices for food and other necessities on the black market reached beyond the means of most Cubans. The solution to control escalating black market prices was desperate: in 1993, the Cuban government legalized hard currency in the state sector, eventually creating a convertible dollar system in 2004 that now competes with the peso system of wages and subsidies. Joint ventures were legalized and tourism was developed and promoted, leading to visible differences between foreign (and privileged local) consumers and Cuban 'workers'.

Yet, contrary to neoliberal projections of Cuban transition, in 1999 the Cuban government initiated another swing towards Guevarian moralism, with a campaign similar to Fidel Castro's 1986 'Rectification' called the 'Battle of Ideas'. Like earlier mobilizations, the goal of this most recent campaign has been to mobilize Cubans to engage in voluntary *and* paid labour to construct clinics and schools, to increase the number of social workers, to eliminate malnutrition, to increase school attendance and the availability of computers, and so on. In addition to welfare measures to counter rising inequalities, there has been a renewed drive to motivate the Cuban people to resist the temptations of profiteering in the black market and to continue the Fight of the Cuban people through self-sacrifice, agricultural production and hard work.

But the Guevarian emphasis on work and sacrifice in the domestic sector increasingly conflicts with market and leisure opportunities in the tourist sector. Richard Feinberg (cited at the beginning of this chapter) intimates this contradiction when he refers to North Americans' use of Cuban golf courses and marina facilities, among other market and leisure opportunities increasingly available to a select few. The paradoxes and inequalities of Cuba's dual economy do not cancel out long-term values and claims to justice that have come to define Cuba as a revolutionary society, however. Feinberg's assumption that all Cubans will be better off once the country succumbs to 'the powerful gravitational pull of geography' ignores past and present inequalities within

Cuba and between Cuba and the United States. His assumption is based on the liberal idea that market opportunities *will* improve the general welfare, 'trickling down' to all Cubans. As I argued at the beginning of this chapter, such predictions for Cuba's temporal and spatial transition are not morally neutral. As Gunnar Myrdal (2002 [1953]: 209) once wrote of the theoretical precursors to neoliberalism: 'To say that most people will have gained after a time and that the losses of those who have not will be negligible involves both interpersonal comparisons of utility and a value judgement'. Recognizing underlying values embedded in the neoliberal model is critical for issues of justice, for if all individuals are seen as equally capable of maximizing their utility (or happiness) in the market, then it is just one more step to disregard and even exacerbate all kinds of inequalities, including those of a more embodied kind:

> If person A as a cripple gets half the utility that the pleasure-wizard person B does from any given level of income, then in the pure distribution problem between A and B the utilitarian would end up giving the pleasure wizard B more income than the cripple A. The cripple would then be doubly worse off: both since he gets less utility from the same level of income, *and* since he will also get less income. Utilitarianism must lead to this thanks to its single-minded concern with maximizing the utility sum. (Sen 2009: 203)

In the neoliberal model, public goods like food, water, education and healthcare should be priced like any other commodity: according to the supply and demand of equally competitive, equally rational consumers. Amartya Sen (2009) argues that such a perspective fails to see the different natural capabilities of individuals: a disabled man may get fewer utilities in the market because his favourite shop has no wheelchair ramps.

In addition to such embodied inequalities, the model neglects 'the accidents of history and geography that endow people with different, and unequal, resources and entitlements, and it offers those who are relatively well-off grounds to obstruct pro-poor government actions as compromising their utility' (Rankin 2004: 198). Indeed, neoliberals claim a 'politics of privileged representation' (Dilley 1992: 2) in accordance with the perspective of a person capable of maximizing his utility in the market. But they are not the only ones to do this. A similar moral straightjacket is imposed by many advocates of the Cuban socialist model. The latter is also based on a singular spatio-temporal trajectory for Cuba, sidestepping the 'multiple "Cubas" on which various intellectual projects and political discourses ... [have been] built' (Hernandez-Reguant 2005: 279). Admittedly, the revolutionary government has promoted and, in some cases, increased the sphere of public goods available to all citizens. But it has also hindered more individualized agencies and identities by defining who gets what, where, how and why. Continuous with the gentlemanly tradition of the first Cuban revolution, those who promote the present 'Battle of Ideas' campaign have the authority to distinguish collective needs from individual desires, favouring the former over the latter. And, at least since the early 1960s – when counter-hegemonic ideals became hegemonic realities – the emphasis on universal needs has coincided with scarcities.

As we shall see in the following chapters, the people of Tuta have been able to live through extreme scarcities, and not just because of the coercive hand of the state. Though to a large extent the national Fight has been transformed into a localized fight for provisions, Tutaños have been able to reproduce themselves both physically and spiritually, in part by re-incorporating transcendental values embedded in the history of Cuba as a revolutionary society. As Antoni Kapcia (2008: 85) argues, a majority of Cubans are 'not "against" but "within"' the Revolution, still identifying with an idea that has long been a defining aspect of different versions of Cuban nationhood.

> The average Cuban is still (often to the surprise and puzzlement of the outsider) remarkably committed to many of the basic ideals and precepts of *la Revolución* and understandably defensive about outside criticisms ... In this sense, every Cuban who remains more or less a part of the whole project of *cubanía revolucionaria* daily echoes the sentiments of W.B. Yeats – 'Tread softly, because you tread on my dreams'. (Kapcia 2000: 268)

Notes

1   See also Radcliffe (2012), Reid-Henry (2012).
2   Feinberg is a former senior director of the US National Security Council's Office of Inter-American Affairs and Professor of International Political Economy at the University of California, San Diego.
3   Cuba was a member of the IMF until 1964, when, under Ernesto 'Che' Guevara's economic plans, it withdrew from the fund, settling its debt of about $25 million (Feinberg 2011: 66).
4   For discussions of 'post-neoliberalism' in geography see, e.g., Castree (2010); Peck *et al.* (2010); Laing (2012).
5   That is, Barnett (2005); Heynen *et al.* (2007); Castree (2006, 2008, 2010); Larner (2003); Leitner *et al.* (2007); Peck (2010).
6   In future publications I hope to use this kind of analysis for neoliberal contexts as well.
7   This citation derives from a 1958 speech by Fidel Castro (cited by Thomas 1971: 874).
8   President Grau's 'government of 100 days' was established after a general strike, when a group of students, workers and radical intellectuals in Havana successfully usurped power from President Carlos Manuel de Cespédes. Despite its short period of existence (Grau was replaced by US-backed Carlos Mendieta after 100 days), the influence of the Grau government led to significant reforms such as the elimination of the Platt Amendment in 1934 and, ultimately, the new Constitution of the Republic of 1940.

9   Simón José Antonio de la Santísima Trinidad Bolívar Palacios y Blanco, better known as Simón Bolívar (1783–1830), was the most famous leader of Latin America's independence from Spain. Among the countries he is revered for liberating are: Venezuela, Peru, Ecuador, Bolivia, Colombia and Panama. Similar to José Martí, Bolívar has reached mythical status, as evidenced by the late Venezuelan President Hugo Chávez's decision to use his name for his Bolivarian Alliance for the Americas (ALBA).

10   The United States occupied Cuba in 1898–1902, 1906–1909, 1912 and 1917–1923.

11   That is, Guerra y Sánchez (1964 [1927]); Aranda (1968); Martín Barrios (1984).

12   That is, Mintz (1985); Comaroff (1996); Howes (1996); Comaroff and Comaroff (1997); Tobin (1999).

13   See Veblen (2007 [1899]).

14   Cienfuegos was a Cuban commander in the rebel army. Loved by all Cubans, his mysterious death in late 1959 has led many – including some of my friends – to question whether or not his death was planned.

15   Debray is a French Marxist intellectual, journalist, government official and academic who fought with Che Guevara in Bolivia.

16   According to Tablada (1990 [1987]: 156), Guevara predicted that the entire globe would eventually be integrated into a 'world socialist pricing system'.

17   The following three paragraphs draw from Helen Yaffe's account of Guevara as president of the Cuban National Bank from 1959 to 1961 (Yaffe 2009: 22–30).

18   Carlos Rafael Rodríguez was a Marxist Cuban politician prior to 1959, who then served under Castro as head of the Institute of Agrarian Reform (1962–1965; Che held this position after) and vice-president of the Politburo (Political Bureau of the Party) until his death in 1997. E.G. Liberman was a Russian economist, René Dumont was a French agronomist (cited earlier) and Charles Bettleheim was a French economist and historian. The latter three were advisors to Fidel Castro.

References

Abel, Christopher and Nissa Torrents (eds). (1986) Introduction. In their *José Martí: Revolutionary Democrat*. London: Athlone Press, pp. 1–15.

Aguilar, Louis E. (1998 [1993]) Cuba 1860–1930. In *Cuba: a Short History*, edited by Leslie Bethell. Cambridge: Cambridge University Press, pp. 21–56.

Al Roy, Carl. (1967) The peasantry in the Cuban revolution, *Review of Politics* 19: 87–99.

Aranda, Sergio. (1968) *La Revolución Agraria de Cuba*. Mexico: Siglo Veintiuno Editores, S.A.

Baber, Willie A. (1987) Conceptual issues in the new economic anthropology: moving beyond the polemic of neo-classical and Marxist economic theory. In *Beyond the New Economic Anthropology*, edited by John Clammer. Basingstoke: Macmillan, pp. 45–77.

Barnett, C. (2005) The consolation of 'neoliberalism', *Geoforum* 36, 1: 7–12.

Boltanski, Luc and Laurent Thévenot. (2006) *On Justification: Economies of Worth* (transl. by Catherine Porter). Princeton: Princeton University Press.

Bourdieu, Pierre. (2005) *The Social Structures of the Economy* (transl. by Chris Turner). Cambridge: Polity Press.

Burke, Timothy. (1996) *Lifebuoy Men, Lux Women: Commodification, Consumption and Cleanliness in Modern Zimbabwe*. Durham: Duke University Press.

Carciofi, Ricardo. (1983) Cuba in the seventies. In *Revolutionary Socialist Development in the Third World*, edited by Gordon White, Robin Murray and Christine White. Sussex: Wheatsheaf Books, pp. 193–233.

Casteñeda, Jorge G. (1998) *Compañero: the Life and Death of Che Guevara*. New York: Vintage Books.

Castree,, Noel (2006). From neoliberalism to neoliberalisation: consolations, confusions and necessary illusions, *Environment and Planning A* 38(1): 1–6.

Castree, Noel. (2008) Neoliberalising nature: processes, effects, and evaluations, *Environment and Planning A* 40(1) 153–73.

Castree, Noel. (2009) The time-space of capitalism, *Time and Society* 18(1): 26–61.

Castree, Noel. (2010) Crisis, continuity and change: neoliberalism, the left and the future of capitalism, *Antipode* 41(s1): 185–213.

Cohen, Anthony P. (2000a) Introduction: discriminating relations – identity, boundary and authenticity. In *Signifying Identities: Anthropological Perspectives on Boundaries and Contested Values*, edited by Anthony P. Cohen. London: Routledge, pp. 1–14.

Cohen, Anthony P. (2000b) Peripheral vision: nationalism, national identity and the objective correlative in Scotland. In *Signifying Identities: Anthropological Perspectives on Boundaries and Contested Values*, edited by Anthony P. Cohen. London: Routledge, pp. 145–69.

Comaroff, Jean. (1996) The empire's old clothes: refashioning the colonial subject. In *Cross-cultural Consumption: Global Markets, Local Realities*, edited by David Howes. London: Routledge, pp. 19–38.

Comaroff, John L. and Jean Comaroff. (1997) *Of Revelation and Revolution*, Vol. 2: *The Dialectics of Modernity on a South African Frontier*. Chicago: University of Chicago Press.

Davis, Bob. (2006) IMF plotting a hardship 'post-Castro' Cuban economy, *Wall Street Journal* 15 Nov., 1–2. http://www.mombu.com/culture/columbia/t-imf-plotting-a-hardship-post-castro-cuban-economy.com (last accessed 16 May 2013).

De la Torre, Miguel A. (2003) *La lucha de Cuba: Religion and Politics on the Streets of Miami*. Berkeley: University of California Press.

Depestre Catony, Leonardo (1987) *Cuba: En Citas 1899–1952*. Havana: Editorial Gente Nova.

Díaz-Brisquets, Sergio and Jorge Pérez-Lopez. (2000) *Conquering Nature: the Environmental Legacy of Socialism*. Pittsburgh: University of Pittsburgh Press.

Dilley, Roy. (1992) Introduction. In *Contesting Markets: Analogies of Ideology, Discourse and Practice*, edited by Roy Dilley. Edinburgh: Edinburgh University Press, pp. 1–26.

Duany, Jorge. (2000) Reconstructing Cubanness: changing discourses of national identity on the island and in the diaspora during the twentieth century. In *Cuba, the Elusive Nation: Interpretations of National Identity*, edited by Damián J. Fernández and Madeline Cámara Betancourt. Gainsville: University Press of Florida, pp. 17–42.

Dumont, Louis. (1986 [1983]) *Essays on Individualism: Modern Ideology in Anthropological Perspective*. Chicago: University of Chicago Press.

Dumont, Rene. (1973 [1970]) *Is Cuba Socialist?* (transl. by Stanley Hochman). London: Deutsch.

Elias, Norbert. (1969) *The Civilizing Process*, Vol. 1: *The History of Manners*. Oxford: Blackwell.

Elias, Norbert. (1982) *The Civilizing Process*, Vol. 2: *State Formation and Civilization*. Oxford: Blackwell.

Evans-Pritchard, Sir Edward Evan. (1962) *Essays in Social Anthropology*. London: Faber and Faber.

Feinberg, Richard E. (2011) *Reaching Out: Cuba's New Economy and the International Response*. Washington, DC: Latin American Initiative at Brookings.

Fernández, Damián J. (2000) *Cuba and the Politics of Passion*. Austin: University of Texas Press.

Foster-Carter, Aiden. (1976) From Rostow to Gunder Frank: conflicting paradigms in the analysis of underdevelopment, *World Development* 4(3): 167–80.

Goizueta-Mimó, Félix. (1972) *Azúcar Amargo Cubano: Monocultivo y Dependencia Económica*. Madrid: Istituto de Sociología y Eesarrollo del Área Iberica.

Guerra, Lilian. (2005) *The Myth of José Martí: Conflicting Nationalisms in Early Twentieth-Century Cuba*. Chapel Hill: University of North Carolina Press.

Guerra y Sánchez, Ramiro. (1964 [1927]) *Sugar and Society in the Caribbean: an Economic History of Cuban Agriculture*. New Haven: Yale University Press.

Guevara, Ernesto Che. (1971) Man and socialism in Cuba. In *Man and Socialism in Cuba: the Great Debate*, edited by Bertram Silverman. New York: Atheneum.

Guha, Ranajit. (1983) *Elementary Aspects of Peasant Insurgency in Colonial India*. Delhi: Oxford University Press.

Hernandez-Reguant, Ariana. (2005) Cuba's alternative geographies, *Journal of Latin American Anthropology* 10(2): 275–313.

Herzfeld, Michael. (2005) *Cultural Intimacy: Social Poetics in the Nation State*, 2nd revised edition. New York: Routledge.

Heynen, Nik, James McCarthy, Scott Prudham and Paul Robbins. (2007) *Neoliberal Environments: False Promises and Unnatural Consequences*. London: Routledge.

Holbraad, Martin. (2000) *Money and need: Havana in the Special Period, Presentation to the Annual Post-Socialism Workshop*, University College, London.

Howes, David. (1996) Introduction: commodities and cultural borders. In *Cross-cultural Consumption: Global Markets, Local Realities*, edited by David Howes. London: Routledge, pp. 1–18.

Hulme, Peter. (2012) Writing on the land: Cuba's literary geography, *Transactions of the Institute of British Geographers* (N.S.) 37(3): 346–58.

Humphrey, Caroline. (1989) 'Janus-faced signs' – the political language of a Soviet minority before Glasnost. In *Social Anthropology and the Politics of Language*, edited by Ralph Grillo. London: Routledge.

Ibarra, Jorge. (1986) Martí and socialism. In *José Martí: Revolutionary Democrat*, edited by Christopher Abel and Nissa Torrents. London: Athlone Press, pp. 82–111.

Kapcia, Antoni. (1986) Cuban populism and the birth of Martí. *José Martí: Revolutionary Democrat*, edited by Christopher Abel and Nissa Torrents. London: Athlone Press, pp. 32–4.

Kapcia, Antoni. (1997) Political and economic reform in Cuba: the significance of Che Guevara. In *La Situación Actual en Cuba: Desafíos y Alternativas*, edited by Mona Rosendahl. Stockholm: Institute of Latin American Studies, Stockholm University.

Kapcia, Antoni. (2000) *Cuba: Island of Dreams*. Oxford and New York: Berg.

Kapcia, Antoni. (2008) *Cuba in Revolution: a History since the Fifties*. London: Reaktion Books.

Knight, Franklin W. (1990) *The Caribbean: the Genesis of a Fragmented Nationalism* (2nd edn). Oxford: Oxford University Press.

Laing, Anna Frances. (2012) Beyond the zeitgeist of 'post-neoliberal' theory in Latin America: the politics of anti-colonial struggles in Bolivia, *Antipode* 44(4): 1051–4.

Larner, Wendy. (2003) Neoliberalism?, *Environment and Plannning D* 21, 4: 509–12

Le Riverend, Julio. (1966) *La República: Dependencia y Revolución*. Havana: Editora Universitaria.

Leitner, Helga and Byron Miller. (2007) Scale and the limitations of ontological debate: a commentary on Marston, Jones and Woodward, *Transactions of the Institute of British Geographers* 32(1): 116–25.

Macphereson, C.B. (1979 [1962]) *The Political Theory of Possessive Individualism. Hobbes to Locke*. Oxford: Oxford University Press.

Martí, Jorge L. (1959) *Cuba: Conciencia y Existencia*. Havana: Editorial Libreria Martí.

Martí, Jose. (1975) *Inside the Monster: Writings on the United States and American Imperialism* (ed. and transl. by Philip S. Foner). New York and London: Monthly Review Press.

Martí, Jose. (1977) *Our America: Writings on Latin America and the Struggle for Cuban Independence*, (ed. and transl. by Philip S. Foner). New York: Monthly Review Press.

Martín Barrios, Adelfo. (1984) Historia política de los campesinos cubanos. In *Historia política de los Campesinos Latinoamericanos*, Vol. 1, edited by Pablo González Casanova. Mexico: Siglo Veinti Uno Editores, pp. 40–92.

Martínez-Alier, Juan. (1977) *Haciendas, Plantations and Collective Farms: Agrarian Class Societies, Cuba and Peru*. London: Cass.

Massey, Doreen. (2011 [2005]) *For Space*. London: Sage Publications.

McClintock, Anne. (1995) *Imperial Leather: Race, Sexuality and Gender in the Colonial Contest*. London: Routledge.

Meso-Lago, Carmelo. (1981) *The Economy of Socialist Cuba*. Albuquerque: University of New Mexico Press.

Mintz, Sidney. (1956) Cañamelar: the subculture of a rural sugar plantation proletariat. In *The People of Puerto Rico*, edited by Julian H. Steward, Robert A. Manners, Elena Padilla Seda, Sidney W. Mintz and Raymond L. Scheele. Urbana: University of Illinois Press, pp. 165–93.

Mintz, Sidney. (1985) *Sweetness and Power*. Boston: Penguin Books.

Montejo, Esteban and Miguel Barnet. (1993 [1968]) *The Autobiography of a Runaway Slave*. London: Macmillan.

Myrdal, Gunnar. (2002 [1953]) *The Political Element in the Development of Economic Theory*. London: Routledge and Kegan Paul.

Ortiz, Fernando. (1993 [n.d.]). *Etnia y Sociedad*. Havana: Editorial de Ciencias Sociales.

Ortiz, Fernando. (1995 [1940]) *Cuban Counterpoint: Tobacco and Sugar*. Durham: Duke University Press.

Peck, Jamie. (2010) *Constructions of Neoliberal Reason*. Oxford: Oxford University Press.

Peck, Jamie, Nik Theodore and Neil Brenner. (2010) Postneoliberalism and its malcontents, *Antipode* 41(s1): 94–116.

Pérez, Louis A., Jr. (1986) Politics, peasants and people of color: the 1912 'race war' in Cuba reconsidered, *Hispanic American Historical Review* 66(3): 509–39.

Pérez, Louis A., Jr. (1995 [1988]) *Cuba: Between Reform and Revolution*. Oxford: Oxford University Press.

Pérez, Louis A., Jr. (1998) *Cuba between Empires 1878–1902*. Pittsburgh: University of Pittsburgh Press.

Portell Vilá, Herminio. (1995 [1939]) *Historia de Cuba*. Havana: Montero.

Radcliffe, Sarah A. (2012) Relating to the land: multiple geographical imaginations and lived-in landscapes, *Transactions of the Institute of British Geographers* (N.S.) 37(3): 359–64.

Rankin, Katherine Neilson. (2004) *The Cultural Politics of Markets: Economic Liberalization and Social Change in Nepal*. London: Pluto Press.

Redfield, Robert. (1956) *Peasant Society and Culture: an Anthropological Approach to Civilization*. Chicago: Chicago University Press.

Reid-Henry, Simon. (2012) Geography and metaphors: a response to 'Writing on the land', *Transactions of the Institute of British Geographers* (N.S.) 37(3): 365–9.

Rostow, Walt Whitman. (1960) *The Stages of Economic Growth: a Non-communist Manifesto*. Cambridge: Cambridge University Press.

Sayer, Andrew. (2000) Moral economy and political economy, *Studies in Political Economy* 61: 79–104.

Sayer, Andrew. (2011) *Why Things Matter to People: Social Science, Values and Ethical Life*. Cambridge: Cambridge University Press.

Scheele, Raymond L. (1956) The prominent families of Puerto Rico. In *The People of Puerto Rico*, edited by Julian H. Steward, Robert A. Manners, Eric R. Wolf, Elena Padilla Seda, Sidney W. Mintz and Raymond L. Scheele. Urbana: University of Illinois Press, pp. 418–62.

Sen, Amartya. (2009) *The Idea of Justice*. London: Penguin Books.

Silverman, Bertram. (1971) Preface. In *Man and Socialism in Cuba: the Great Debate*, edited by Bertram Silverman. New York: Atheneum, pp. 1–28.

Steward, Julian H. (1956) Introduction. In *The People of Puerto Rico*, edited by Julian Steward, Robert A. Manners, Eric R. Wolf, Elena Padilla Seda, Sidney W. Mintz and Raymond L. Scheele. Urbana: University of Illinois Press, pp. 1–28.

Suárez Salazar, Luis. (2000) *El Siglo XXI: Posibilidades y Desafíos para la Revolución Cubana*. Havana: Editorial de Ciencias Sociales.

Sutton, David. (2001) *Remembrance of Repasts: an Anthropology of Food and Memories*. Oxford: Berg.

Szulc, Tad. (1989 [1986]) *Fidel: a Critical Portrait*. Dunton Green: Coronet Books.

Tablada, Carlos. (1990 [1987]) *Che Guevara. Economics and Politics in the Transition to Socialism*. New York: Pathfinder.

Thomas, Hugh. (1971) *Cuba: or the Pursuit of Freedom*. London: Eyre and Spottiswoode.

Thomas, Hugh. (1998 [1993]) 1750–1860. In *Cuba: a Short History*, edited by Leslie Bethell. Cambridge: Cambridge University Press, pp. 21–56.

Tobin, Beth Fowkes. (1999) *Picturing Imperial Power: Colonial Subjects in Eighteenth-century British Painting*. Durham: Duke University Press.

Tsing, Anna. (2000) The global situation, *Cultural Anthropology* 15(3): 327–60.

Valdés, Orlando. (2003) *Historia de la Reforma Agraria en Cuba*. Havana: Editorial Ciencias.

Veblen, Thorstein. (2007 [1899]). *Theory of the Leisure Class*. Middlesex: Echo Library.

Wilk, Richard (ed.). (1999) Real Belizean food: building local identity in the trans-national Caribbean, *American Anthropologist* 101(2): 244–55.

Wilk, Richard (2006) *Home Cooking in the Global Village: Caribbean Food from Buccaneers to Ecotourists*. Oxford: Berg.

Williams, Colin C. (2005) *A Commodified World? Mapping the Limits of Capitalism*. London: Zed Books.

Williams, Raymond. (1990 [1973]) *The Country and the City*. New York: Oxford University Press.

Wolf, Eric. (1969) *Peasant Wars of the Twentieth Century*. New York: Harper Books.

Yaffe, Helen. (2009) *Che Guevara: the Economics of Revolution*. London: Palgrave Macmillan.

Chapter Three
Scarcities, Uneven Access and Local Narratives of Consumption

'The wine is sour, but it is *our* wine.' (Martí 1975)

In the previous chapter I outlined the historical and political economic contours of Cuban socialism as a 'philosophy of worth' (Boltanski and Thévenot 2006), which has, over time, been strengthened by its opposition to another value system tied to the global market economy. This background information is essential for my primary argument in this book, which is that Tutaños and Tutaño households are materially and morally emplaced in particular ways vis-á-vis national and global imaginaries of economy. Power geometries relate Tutaños as individuals to wider scales such as the nation state or to global market networks, and these relations are moral as well as political economic. One way to uncover how such scalar politics work in practice is through empirical evidence of how Tutaños justify scarcities and differential access to food, markets and productive inputs. Chapters 3–6 of this book are geared towards the discovery of these material, moral and spatial processes of justification through ethnographic data from my own studies in the field. I uncover how Tutaños adopt or change the cultural politics and policies of the nation state, which are explained in turn, to rationalize, contest and sometimes exacerbate inequalities of access to food and other privileges.

As in all places, individuals in Tuta (Figure 3.1) shift between different 'orders of worth' (Boltanski and Thévenot 2006) – some of which may be closer to ideal norms than others. The national moral economy influences

Everyday Moral Economies: Food, Politics and Scale in Cuba, First Edition. Marisa Wilson.
© 2014 John Wiley & Sons, Ltd. Published 2014 by John Wiley & Sons, Ltd.

Figure 3.1   A picture of Tuta from above. Source: author.

rather than determines how Tutaños justify their differential positionings as consumers, buyers/sellers and producers of food products. In practice, it is often the case that actors 'remai[n] true to a consistent set of requirements' (Boltanski and Thévenot 2006: 15), which in Cuba derive from long-term values that developed in relation to a contrasting value system just 90 miles away. The work of social scientists is to spell out these 'set[s] of require-ments', not by further empowering the Leviathans that already exist, but by explaining internal arrangements that allow for the emergence of dominant (if not entirely determining) systems of value:

> Our descriptive enterprise, conducted from within each world, requires our reader to suspend the critical outlook that results … from familiarity with several different worlds, and to plunge into each world in turn as one would do in a situation in which the sincerity of one's adherence to principles would be a condition of the justification of one's action. (Boltanski and Thévenot 2006: 136)

Unless the analyst suspends his or her critical eye, understanding the nor-mative conditionalities through which material conditions are reproduced is impossible. If I had not done so in Cuba, I would have got no further than seeing Cuban life from a tourist's perspective, searching for the best pizza at the best price.

But the importance of describing the contours of everyday life should not eclipse ongoing and potentially life-changing events, lived through and shaped by certain individuals (i.e. serendipitous interactions between Cubans and myself). If 'event-spaces'[1] are never predetermined, then all cultures contain 'the embryo of another cultural order' (Sahlins 1976: 43). Moreover, if the focus is on static description more than analysis, one may risk reifying a homogeneous spatio-temporal 'community' that corresponds to an unchanging system of values. In an attempt to avoid such sociological pitfalls, in the next four chapters I shift constantly from an epistemological description of the Cuban moral economy as a *normative* 'project of community' (Gibson-Graham 1996: xxxii) to a more substantive analysis of *empirical* relations between individuals that may or may not conform to its ideal representations. Like all projects for social cohesion, it is clear that loopholes exist between the formal model of Cuban socialism and the substantive realities of daily existence. My aim in this book is to reach an analytical balance between describing ideal principles of justice and value in Cuba and uncovering how such principles are adopted or counteracted by particular people in the difficult (i.e. non-ideal) conditions of their everyday lives.

In this chapter and chapter 4, I explain how local narratives of consumption draw from representations of the national moral economy, and show how consumer narratives are used to justify or contest everyday material realities. Similarly, in chapters 5 and 6, I reveal how the scalar politics of Cuban nationhood influences moral, material and spatial processes of exchange/distribution and production, respectively. The final chapter places all this normative and empirical data in the light of wider obstacles and possibilities for social and environmental justice in Cuba. The portrait of these obstacles and possibilities begins to take shape in this chapter. In the first part (divided into two subsections), I explain ideal/official conceptualizations of the national food system, comparing these to real-life experiences and rationalizations of uneven access. In the second part (with three subsections), I incorporate other ethnographic examples to show how Tutaños account for scarcities and inequalities with a unique sense of humour and with locally acceptable ways of expressing desire. Such narratives of consumption resemble the ways Tutaños discussed hunger, nourishment and need, particularly as remembered during the most severe 'Special Period' of scarcities in the early to mid 1990s. The latter cultural idioms will be discussed in chapter 4 in the light of official justifications for recent changes to the socialist contract between state and citizens.

The ideal of national re-distribution and the reality of uneven access

Gaps in the national oikos

The Cuban state ideally maintains a reciprocal relationship with its citizens. The latter offer their labour power for the production of collective property, while the former is legitimized by providing security and minimum needs for all citizens. As emphasized, food in this system is treated as a good that can be universally obtained with token pesos rather than a commodity with prices set by supply and demand. In an article in the communist paper, *Granma*, in June 2008, the Cuban government and Party re-affirmed their commitment to the people by insisting that farmers should be more inclined towards their 'natural market', a re-distributive institution called the Acopio, than the farmers' market or any other market with prices determined by global exchange values (*Granma* 2008: 2).

As an institution that regulates economic transfers and values within the nation, the Acopio is comparable to Aristotle's 'natural economy': the household or *oikos*. In 1925, Alexander V. Chayanov noted that prices for goods are absent in economies based on the Aristotelian idea of the *oikos*, of which he claimed there were three: the household, the peasant economy and the communist economy (Chayanov 1966 [1925]: 25). In the household and peasant economies, the economy is a bounded unit, in which production and consumption work towards the reproduction of the unit. The 'natural economy' of the Cuban nation, along with its 'natural market', the Acopio, is based on Chayanov's third type of economy: the communist nation. Like the household and peasant economies, the communist nation is treated as a closed system, where industrial and agricultural outputs of workers (and 'peasant-workers') are redirected to the centre, to then be re-distributed to individual citizens according to need. Accordingly, the Cuban government prioritizes need and other values such as collective work effort, which remain valid within the territory of the nation state.

Even the exchange values of commodities in Cuban markets, which theoretically work according to 'global' laws of supply and demand, are not impermeable to the state's value system. In addition to the periodic implementation of price controls, the state-run Acopio enterprise can take any commodity off the farmers' market if there is a shortfall in rationed items (Wright 2008: 22) or if the state sees a need to sell a certain food item at a 'just' price. As one *campesino* (farmer) who was selling avocados and black beans at the farmers' market in Tuta told me: 'I do not have any plantains today. The state market needs plantains, so no one in the market has plantains'. Instead of stockpiling produce to increase prices (as in commodity futures markets), the Cuban state hoards foodstuffs in order to

Figure 3.2   A vendor measures black beans at the farmers' market. Source: author.

decrease prices: 'The state … flood[s] the market to lower the price of any commodity whose price was too high on the market' (Wright 2008: 245). But as in other (post)socialist places, there are often not enough state-subsidized goods to satisfy demand. People like my Cuban 'family' have to rely on higher-priced outlets for food, such as the farmers' market (Figure 3.2), which has been dubbed the *'Bandido del Rio Frio* (Cold River Bandit)' after a popular Brazilian soap opera, because prices charged in this market are seen as 'abusive'.

Neoliberal economists (e.g. Marshall 1998; Ritter 1998; Álvarez 2004a) claim that 'sticky' or consistently high prices (Álvarez 2004a: 6) in Cuban markets are due to political obstacles which impede the 'natural' workings of the market (again note the language of naturalism). But, as Chayanov's writings suggest, the issue of prices is much more complicated, as it relates to the degree to which an economy works as a bounded, self-sustaining unit, as visualized at different scales.

We have seen that the Cuban state has had to open up its borders to out-side exchange values which compete with the use values it measures and controls within the national *oikos*. Even so, the presence of the market has not led to a uniform transition to capitalist market relations. As I will argue with empirical examples, economic transactions made between smallholders and their clients, or between intermediaries and consumers, are not entirely based on values set by market prices, according to which individual maximization sets the relationship between supply and demand. Rather, values that shape provisioning processes in Tuta are tied to particular relations between inside and outside forces which come to the fore as economic practices unfold in everyday life.

Because rural areas like Tuta are places where the 'market principle' has not yet fully penetrated the 'market place' (Herskovits 1962: viii; Bohannan and Dalton 1962: 1–2), it is *especially* evident that the market (capitalist) concept, or the idea of an amalgamation of individuals who always act in pursuit of their own interests, is an insufficient explanation for ordinary economic geographies in Tuta. To rely solely on the metaphor of the market ignores other kinds of internal exchanges that take place within and between market settings (Herskovits 1962: xiv), including (but not limited to) the regulatory power of the state.

Given contradictions in Tuta between territorial controls and external influences, it is clear that markets in Tuta (like all other markets) are not 'neutral meeting grounds for equivalent exchanges among economic equals … [but] … arenas of encounter and conflict' (Wolf 2001: 283). Though price regulations and other state controls partly contribute to Cuba's ideal of 'food for all', they have also led to stark contrasts between prices and products available in state-subsidized markets and those offered at world market prices. Normatively, however, the 'natural market' for the nation's needs (the Acopio) is valued as infinitely superior to all other market(s) in the country, in line with a value system that emphasizes use values over exchange values. Like Chayanov who relied on an underlying notion of 'minimum needs' for the household, so people who uphold the Cuban moral economic model value the *canasta básica* (basic food basket) more than consumer demand.

Accordingly, the 'socialist food discourse' (Pertierra 2007: 100) in Cuba emphasizes physiological needs more than cultural desires, calories and levels of nutrition more than taste or local preferences. Food rations allocated to citizens, portions served in workers' canteens, and crops produced in the agricultural sector are all listed in detail, with weights and volumes specified in ration books, menus, newspapers and governmental announcements. Food allocation by the Cuban state is determined and evaluated principally by nutritional criteria, and the nutritional status of the Cuban population is a principal way of measuring the progress of the Cuban Revolution (Pertierra 2007: 61).

But while quantitative measurements work as 'visual representations of order and efficiency', they often ignore 'illegible' aspects of local systems of value (Scott 1998: 223, 1–2). Over 80 years ago in her study of Bantu morals of food, anthropologist Audrey Richards made a similar point, urging economists, scientists and other primary agents of food policy to take into account '[h]uman emotions, which are from their very nature passionate, complex and infinitely various, [and which] rebel against the scientists' classifying hand' (Richards 1932: 116). As Richards argued, localized understandings of food consumption, exchange/distribution and production are inseparable from what she called 'nutritive relationships' (Richards 1932: 116), or historical, political economic and socio-cultural formations of food networks and values, which are often neglected in official contexts.

Just as neoliberal economists overlook long-term revolutionary values in Cuba, so strict defenders of the national moral economy fail to account for local processes of valuation. Indeed, universalistic models to satisfy Cuban needs leave no space for cultural, spatial and historical frameworks through which material culture is used and symbolized in ordinary life.

> [I]f there is a universal property, it is that agents are not universal, because their properties, and in particular their preferences and tastes, are the product of their positioning and movements within social space, and hence of collective and individual history. (Bourdieu 2005: 211)

Everyday consumer patterns in Tuta have been tied not only to national and socialist norms, but also to global circuits of value, measured in terms of prices but also in terms of quality. Tutaños contrasted the increasing presence of outside markets in present-day Cuba with an earlier time in their history when commodities entering and exiting the market did not have to compete with collective goods in the re-distributive economy. As one Tutaño said: 'I am hoping that trade with the United States continues to increase.[2] People have always wanted US goods. They are better than what the state offers'. As suggested in the previous chapter, 'dependence upon imported consumer goods [in Cuba] was established as part of "local" culture long before the introduction of socialism' (Pertierra 2007: 48). Local memories of 'the times of capitalism' partially de-legitimize the scalar and cultural politics of the Cuban nation state, and cyclical shifts between state and market continue to affect when and how Cubans enter into market spaces. As in 1950s to 1980s China, in Cuba periods of 'temporary' openings to the market have occurred at least since the 1970s, and the present period may be no exception.

Contrary to neoliberal transitologists, however, some Tutaños with whom I interacted actually preferred the universal distribution of lower-quality goods to commodities bought in the market. When justifying their preferences for social over individualistic forms of provisioning, many quoted José Martí: 'The wine is sour, but it is *our* wine', an expression he wrote during the first revolutionary struggle to encourage Cubans to 'resist' the temptations of commodities from the imperialist 'monster' of the north (Martí

Figure 3.3   Men helping to build a house. Source: author.

1975). One case in point was when I offered to buy comparatively expensive rum (Havana Club) for my Cuban father, Jorge, who insisted upon bringing home low-quality rum known colloquially (and appropriately) as '*gasolina*'. When I asked him why he preferred what he admitted to be lower-quality rum, his response was revealing:

> This rum is cheap, it comes from *Them* [the state]. Everyone can buy it. It is good because of this. And everyone got a bit of it at our neighbor's house [all the men went to the neighbour's to help construct his house; Figure 3.3], but I asked for a small bottle for tonight and worked harder. Now I know my neighbor will come and work for us when we need it. (Wilson 2012a: 286)

In this book, we will encounter other instances where 'food for all' (or perhaps, more appropriately, 'food for work') is seen as better than individual manifestations of self-interest, whether the choice for asceticism is inspired by cultural factors or by the visible presence of coercive power. Such moral aspects to economic life call into question Veblenian assumptions of the 'universal desirability of "luxury" goods' (Wilson 2012a: 286) for, as Andrew Sayer (2003: 347) argues 'consuming because others consume may not necessarily be driven by a desire to demonstrate superiority over them but to achieve equality with them'. Yet while Tutaño consumption was motivated by social over individual interests in some contexts, personal desires were also relevant in everyday life, a point I will illustrate in the next section.

Collective needs versus individual desires

Despite social influences (or pressures) that inhibit personal satisfaction, many people with whom I interacted in Tuta were fully aware of differences in quality between goods provided by the state to satisfy needs and items they saw as more desirable. Indeed, because the government is more interested in the nutrition of Cuban bodies than the preferences of Cuban palates, many items on offer in state outlets do not appeal to Cubans. When carrots were being sold at low cost at the state fair (*feria*) in Tuta, which sells surplus items from state storage facilities several times a year (though it is supposed to be monthly), I heard one man make the comment: 'Cubans are not rabbits! No Cuban likes carrots!'

Another example of the value of individual tastes over collective re-distribution occurred when a Tutaño *campesino* called dietary substitutes for scarce items, such as *picadillo de soya* (soya mince) an 'imposition' on individual consumers. As we shall see in reference to Cuban irony, many people in Tuta made jokes about state provisions, especially in regard to what they called 'OCNIs' or *objecto comestible no identificados* (non-identified edible objects; Díaz Vázquez 2000: 51; cited by Álvarez 2004b: 5–6), an acronym used to describe substitute meat products such as *picadillo de soya*. I was told that the reason there is a scarcity of beef in *picadillo de soya* is that many people who work in meat factories 'grab' their share of beef, or take state property from their workplace (I return to this idea of 'grabbing' in chapter 5).

As with coffee, dairy, higher-quality seafood and produce, the state regulates the sale and distribution of beef, the most highly valued meat in Cuba. During the first years of the Revolution, most cattle ranches in Cuba were converted into dairy farms to further the goal of providing all Cuban children under 5 years of age with calcium. Despite a lack of fodder, the revolutionary government decided it was best to import cattle of the Holstein variety, which are first-class milk producers and now the most common breed for the production of milk for UK and US supermarkets. Prioritizing milk-producing cows to satisfy the needs of Cuban children left a gap in previously large supplies of beef for Cuban consumption. The state began to monopolize supplies as early as the 1960s, paying cattle farmers much less than they had previously received. When Cuban ranchers reacted by threatening to stop producing beef, all the pasture areas were nationalized, all ears of cattle tagged (interview with history professor, Agrarian University of Havana, 19 May 2007). As one informant said, pointing to a group of scattered cows chewing cud on a vast area of state land, 'All cows are Fidel's cows'.

During the 2-year period I spent visiting Tuta to conduct fieldwork, prices for beef on the black market rose from 25 pesos per kilogram to 38, and then beef ceased to be available at all (though in 2011 beef mince was available through CUC stores at largely inhibitive prices). Even if a household owns cattle (or, more likely, a single bull or cow), the state severely punishes their slaughter with long-term prison sentences. According to an article posted on www.cubanet.org, a site created by an anonymous journalist from Havana, a man who recently slaughtered his own cow committed suicide after officials broke into his house and placed him under arrest. His last words were: 'I'm not going to jail; you want to put me away for many years'.[3] Such legal restrictions on home cattle slaughter are still in place as of the time of writing.

As illustrated by this example and those that follow, the scalar politics of alimentary need in Cuba competes with localized values of food that have developed in the place of rural Cuba. With a diet largely rooted in Spanish and African eating habits, most Cubans do not like soya products, preferring meat over other kinds of protein. Beef is not only valued because it is seen as nutritious ('¡alimenta!' [it nourishes!]), but also because, in Latin American contexts at least,[4] it symbolizes strength. A warm meal in Cuba (usually the same prepared food eaten over the course of lunch and dinner) must include, at the very least, some kind of meat or fish. When I visited my friend in the hospital and asked her how the food was, she said 'I don't eat anything here. Look, I am getting as thin as a rail!' About 20 minutes later a tray arrived, with two fried eggs, rice, beans and some overcooked vegetables. As I continued my conversation, my friend proceeded to eat the entire meal. I could not help but say: 'Well, it looks like they have finally fed you!' She responded: 'Fed me? This is not a meal [no es comida]! Did you see it? There was no meat on my plate!' It seems that other Cubans would have been so offended that they would not have even touched the plate, for according to Julia Wright's interview with a Cuban nutritionist: 'The people's way of resisting is to refuse to eat what may be offered to them, and so the state has learned not to offer what may be rejected' (Wright 2008: 236).

As with meals served in state institutions (e.g. work units, hospitals, universities), Tutaños expressed frustrations with what was on offer in state markets, as selection in these markets is related to state controls over production. In a heated conversation, one infuriated Tutaño campesino, Angelo, contrasted the state market with the 'real' market for demand. 'There is an abyss between the two markets: one of necessities, the other of tastes!' Angelo was frustrated that the state's model for production did not allow for feedback between the two 'markets'. He complained about the kinds of seeds available through the seed enterprises, the only legal outlets where seeds may be purchased. According to Angelo, one can only plant varieties authorized by officials. The resulting produce 'satisfies neither necessity nor consumer taste'. He criticized the state for 'car[ing] more about quintales[5] than quality'.

But while Tutaños live in a world where both state and market values coexist and often contradict one another, some market spaces are entirely restricted to 'outside' consumers. For instance, Cuban consumers must compete with tourist and export markets for tropical fruits, especially citrus. An enterprise specifically dedicated to the production and packaging of citrus products is located near Tuta. When I asked a Tutaño with knowledge of the enterprise about whether hurricanes had destroyed citrus production that year – it was the season for citrus and I did not notice anyone selling oranges or limes in the market – he told me that it was a 'political decision' to direct all citrus production and manufacturing to export and tourism, rather than sell these products in the low-priced state markets. 'The people in Tuta are disgusted about this!'

Many other people were aware of differences in availability and quality of commodities from tourist and external markets as opposed to goods they could obtain from the state. With this knowledge, most Tutaños tried hard to satisfy my tastes. As an outsider, I was expected to desire a wider variety of higher-quality foodstuffs than those on offer to Tutaños. Discussing this subject, one woman explained to me: 'You and everyone *afuera* are used to having all kinds of goods, at any time of the year. We only have access to tomatoes, lettuce, and other healthy items when they are in season'. With reference to a period when few salads and vegetables were available in the farmers' market or state markets, another Tutaño exclaimed to me one day: 'It is impossible to maintain a diet here! In the market, there are only avocados! This is the only salad item available, and the fruit is too expensive. One must take produce destined for the tourist market ... the family has to eat!'

The more information people in Tuta have about what people eat in other places the more apparent are differences in access between local markets and the commodities available to outsiders. While the Cuban state sees tourism as benefiting all Cubans, many Tutaños felt left out. Frustrations were most evident regarding products such as beef which, as we have seen, were monopolized by the Revolution early on. Like beef, quality fish and seafood are monopolized by the state for the tourist and export markets. Very rarely do these products become available via state outlets, and at a much lower quality and quantity than those sold in the tourist market. Moreover, Tutaños are wary of black-market sales unless they know the *particular* (vendor) selling the fish or crab (see chapter 5). Blue crab, called 'the Cubans' lobster', was available, though I tried it only once and it made me ill. Lobster was not available in Tuta, except very rarely on the black market.[6]

When a *paladar* opened just 1 kilometre outside Tuta serving lobster and beef in pesos, it seemed that problems of access were improving. The restaurant was short lived, however. The closure of this *paladar*, which was primarily for the benefit of Tutaño consumers, signified the growing divide between food for domestic consumption and food for tourists. As one friend exclaimed, counteracting the cultural politics of collective ownership: '¡Los turistas son los dueños aquí!' (the tourists are the owners here!).

Local narratives of consumption and the Fight

> The ideal society is not outside of the real society; it is part of it. ... For a
> society is not made up merely of the mass of individuals who compose it, the
> ground which they occupy, the things which they use and the movements
> which they perform, but above all is the idea which it forms of itself. (Durkheim
> 1912, cited by Sahlins 1976: 112–13)

Being a luchador

Scarcities and the uneven accessibility of desirable food are clearly visible to
people in Tuta, and these adverse conditions are discussed on a regular
basis. As anthropologist Anna Cristina Pertierra has recently argued, in
Cuba '*everybody* complains about shortages and high prices ... and nobody
can be blamed for doing so. To bemoan the absence of goods need not
imply a defense or a critique of socialist planning policies' (Pertierra 2011:
106; emphasis in original). Though women are most often the cooks in the
family, men, women and children alike must engage in the 'fight for provisions'. When greeting another on the street, it is common to hear the
response: '*Estoy en la lucha ... de provisiones*' (I am in the fight for provisions)
or just '*Estoy en la lucha*' (I am in the struggle).

As indicated in the introduction, words like *lucha* have more than one meaning: they not only point to the scalar politics of the Cuban Revolution, but are
also politically neutral ways of expressing exasperation with the daily difficulties
involved in provisioning for the household. Common expressions like '*lucha de
provisiones*' work as buffers between individual frustrations with the way everyday life in Cuba *is* and moral ideals about how it *ought* to be. The unisemic
reference to *Lucha*, the normative framework that exists above individual experience, may become polysemic – or have multiple references – in the act of
utterance, in the individual performance: '[T]o describe life as a struggle is to
make a wry commentary upon the transition that many citizens made from
struggling *for* Cuban socialism to struggling *with* Cuban socialism' (Pertierra
2011: 98, in reference to Del Real and Pertierra 2008). Thus while the word
'*lucha*' may have two different meanings – one corresponding to the 'is' and the
other to the 'ought' – we must consider the way this distinction may be complicated in actual performance, or localized 'event-spaces'.

Though Cubans' use of words like *lucha* indicates an awareness of the
contradiction between national values and everyday difficulties, this contradiction is partly reconciled by the multiple logics and identities of
individuals, which come to the fore in different contexts. In reference to one
woman's account, Pertierra notes:

> [She]...never seemed uncomfortable about distinguishing between her
> general pride in being 'a Cuban woman', supporting the Revolution and its

achievements, and yet criticizing the shortcomings of the Special Period that she observed had obliged her to struggle to provide good food for the household and to rebuild the house with years of economic sacrifice. (Pertierra 2007: 25)

The contradiction between economic sacrifice for the 'community' (whether defined in terms of the nation, the neighbourhood or the household) and personal frustrations with provisioning is often reconciled by a normative logic that ties economic sacrifice to the daily fight of *all* Cubans. The spirit of self-sacrifice, which we have seen to be a primary value of Cuban nationalism, has now become inseparable from what Cuban sociologist Luis Suárez Salazar refers to as the *heroismo cotidiano* (daily heroism; Suárez Salazar 2000: 292) of all Cubans engaged in the 'fight for provisions' and other daily difficulties. This value of resisting difficult conditions is not only endorsed by the Cuban government, but is present at more localized levels. It is a primary element not only of Cuban national identity but also of local identities. For instance, when I told Tutaños that I travelled in Cuba by lorry (*camión*, the least expensive and most uncomfortable way to travel) rather than by car (the most expensive way to travel), many exclaimed 'You *are* Cuban!' They assumed that, as a 'rich' foreigner, I would have chosen to travel in luxury rather than experience the frustrations and discomfort associated with an unreliable and crowded public transport system.

On another occasion, while conducting research for the Agrarian University of Havana, I chose to eat a nearly tasteless peso pizza for lunch with fellow Cuban students and professors rather than dine at the special cafeteria reserved *exclusively* for foreign students, where more desirable foods were served. Upon taking my first bite, I looked up and saw that nearly all the Cuban students and professors standing in the long line to enter the standard Cuban cafeteria were staring at me. When I inquired about their strange attention, a friend told me, 'They are surprised to see a foreigner eating low-quality pizza when you could have had something better at the cafeteria for foreigners. ... You really must be Cuban!'

From the very beginning of the Revolution to the present, the official way to win the 'war of the people', the battle against imperialism symbolized most powerfully by US sanctions, is to 'resist' (material desires, deprivation, etc.) and to 'fight'. One cannot travel anywhere in Cuba without seeing billboards that contain one or more of these words. Such ideas are linked to a pervasive optimism underlying the Cuban teleology: that individual efforts today will reap collective rewards tomorrow. As Fidel has repeatedly said: '*Luchar por una utopía es, en parte construirla*' (To fight for a utopia is, in part, to construct it; quoted by Suárez Salazar 2000: 363). Local idioms I heard on a daily basis are counterparts to Fidel's optimism: for instance '*Mañana sera mejor*' (Tomorrow will be better); '*Un mundo mejor es posible*' (A better world is possible); '*No hay mal que por bien no venga*' and (Bad only happens for good; my translations).

The revolutionary process, the cornerstone of the national moral economy in Cuba, is characterized by a symbiotic relationship between workers and the revolutionary government. According to this line of thought, working hard, not losing faith and being persistent are central revolutionary values for all Cuban people, which *will* lead to both material and moral rewards. In regard to the latter, Tutaños have proudly shown me certificates of merit from their workplaces, the highest being the title of *Maestro de Oficios* (Master of Office). And contrary to the claim that this kind of revolutionary devotion is on the wane amongst the younger generation, many young Tutaños (aged 18–35) expressed to me, with almost a quixotic sense of duty, the satisfaction they acquired from working hard at their jobs. José, for example, said he never tells people he is tired even if he has worked his two jobs (one state, one private) for 48 hours with little rest. When I asked why I rarely hear him complain about being tired or say he does not want to do something required of him, José simply responded that to say these things would give the impression that he is 'afraid of work', which he is not. His response recalls part of a speech delivered by Fidel Castro during the 'Rectification' campaign in November 1987: 'Our working people are not afraid of anything. Our working people are not afraid of any effort. Our working and revolutionary people are not afraid of the rigor of work' (cited by Tablada 1990 [1987]: 28).

Though such examples of Cubans quoting Fidel directly may indicate political coercion or submission, other cultural idioms of the daily 'fight' suggest that Cuba's scalar politics have entered into localized standards of value. For instance, perhaps the best compliment one can give another person in Tuta is to describe them as a '*luchador/a*' (fighter). I often heard this expression used in praise, not specifically of a person's revolutionary merits, but rather for being a person who confronts hardships with valour. Likewise, the ability to 'resist' is a revolutionary value that allows people to justify everyday difficulties. Quite a few of my friends in Tuta may still agree with a man from Mayarí (eastern Cuba) who was interviewed by a Cuban-American sociologist in the early 1960s: 'Fidel never lied to us about that. He said there would be times when there would not be enough and times when we would have all we want. Fidel never lied' (Yglesias 1968: 106). The ability to resist during hard times was a quality Fidel Castro identified with the Cuban people from the very beginning. Just a week after his 'triumph', Castro conceded to a crowd of eager habaneros:

> I think that this is a decisive moment of our history: the tyranny has been overthrown. Our happiness is immense. But yet, there still remains much to be done. We cannot fool ourselves, thinking of a future when everything will be easy; maybe in the future everything will be even more difficult. (Castro Ruz 2004 [1959]; my translation)

While one may argue that Cubans may no longer have such values when Fidel Castro is no longer in the picture, it is still likely that Fidel will remain a powerful symbol of nationalism and revolutionary strength in Cuba, comparable, at least for some, to José Martí and Ernesto Che Guevara.

Like Fidel, 'good' people in Tuta are considered *luchadores*, as noted above. And, paradoxically, being a 'fighter' actually enables Tutaños to come to terms with how hard their lives are. Perhaps the most common expression I heard on a daily basis in Tuta was '*no es fácil*' (it is not easy), as other anthropologists of Cuba have noted (e.g. Holgado Fernández 2000). Most Tutaños (especially women) describe their lives as 'agitated', as there is always stress involved in procuring the next family meal. Indeed, it seems that all Cubans are aware (even government officials) of the contradiction between the moral drive to resist and frustrations with the way the system is working on the ground. Yet the articulation of the positive and negative aspects of the Revolution is cultural rather than economically rational. While most Tutaños complain about present scarcities (or, perhaps more frequently, the inability to afford items that are available), the same people use revolutionary narratives in their evaluations of people who deal best with the situation. Daily grumbling about life is how many Tutaños identify themselves as 'fighters'; ironically, talking about difficulties behind closed doors often means one is continuing to 'resist' in public.

Being a *luchador* in the local context does not always mean that one follows officially permitted rules, however. Some anthropologists of Cuba (Berg 2004: 48; Roland 2006: 156) have even argued that people use the term '*lucha*' to describe illegal activity:

> The popularly invoked term *lucha* – which can take either the noun or verb form of 'struggle' or 'fight' – encompasses [many] means of surviving in the Special Period. … The irony in the everyday usage of the term … is that it often involves either minor or major illegalities within the Revolution's moralistic system. (Roland 2006: 156)

One should use caution, however, when making parallel associations between legal and moral rules of custom.[7] As Max Gluckman (1965 [1955]) argued for the Lozi of Northern Rhodesia, so in all societies, there *is* a difference between legal rights based on laws enforced in the courts and moral rights or 'the pressure of public opinion, individual conscience and social reciprocity' (Fortes 1969: 237). Legal rules may be broken in everyday life, but illegally licit[8] activity may be distinguished from unacceptable illegal activity in everyday moral economies, as I argue in chapter 5. As used in the moral economy of Tuta, revolutionary ideas like '*la lucha*' may incorporate both legal and illegal activity. But as we shall see later for illegally licit exchanges, grievances about present challenges for the household economy may actually reinforce rather than undermine values set by the national Leviathan. It is in this light – one that incorporates *both* the difficulties and contradictions of everyday material realities *and* the historical values and symbols embedded in Cuban society – that Cuban ironies of consumption should be analysed.

Scarcity and Cuban irony

When I arbitrarily asked one of my Tutaño friends what she thought to be the three most important things in Cuba, she responded, with incremental emphasis: '1. food, 2. Food and 3. FOOD!' Laughing, she added that each corresponds to 'breakfast, lunch and dinner'. With hindsight, I realize that Claudia's comment drew on a common joke in Cuba, to which I shall soon return.

During fieldwork I was surprised to find that even people considered to be closely allied to the Revolution – the *jefes* (state managers) of state enterprises, for example – spoke of food scarcities in ways that could be interpreted as quite anti-revolutionary. One of my informants, Ricardo, a *jefe* of a state dairy farm and general secretary of his Communist Party work nucleus, bewildered me in such a way. When I asked him a simple question: 'What was the biggest hurricane before Ivan?', remembering the destruction this tempest had caused which postponed my second visit to Cuba in 2004, he responded: 'The hurricane of 1959! We lost all our meat, we lost all our crops, the electricity went out, and it hasn't been the same since!' At this, friends and family sitting on the veranda of his sister-in-law's home burst into laughter and I joined in, though slightly confused. The year marking the success of the Revolution was referred to as a moment of destruction rather than its usual depiction by locals and in official texts as 'the triumph'. On another occasion, whilst watching the title sequence of the daily *Noticiero Nacional de la Televisión Cubana* (National News of Cuban Television, abbreviated to NTV in the opening sequence), a former communist leader, Miguel, asked me whether I knew what 'NTV' meant. He did so with a twinkle in his eye. I told him the right answer and he contradicted me, 'No! It does not mean *Noticiero Nacional de la Televisión Cubana*, it means *No Tenemos Viandas* (we have no tubers/root vegetables)!' As tubers and root vegetables are mainstays in the Cuban diet, this joke seemed to be a direct attack on the state's distribution system.

Like the analysis of all forms of representation, that of humour should not start from the assumption that language is a weapon to be used as an outcry against a power structure. Indeed, as Dow and Lixfeld argue, 'the standard political labels of "totalitarian" or "democratic" not only fail to adequately identify the defining conditions for political humour, but obscure the relationship between genre and context' (cited by Stein 1989: 90). Jokes of a political nature must be analysed both within the context in which the joke is uttered and in terms of wider socio-spatial relations. In context, jokes may draw on continuity as well as critique.

Figure 3.4   Two young Tutaños having a chat in a place far from the centre of town. Source: author.

As my fieldwork progressed, I was to find that jokes like Ricardo's are very common in private spaces in Tuta. Some areas of the house, especially those furthest from the 'ears' of the street, are 'off-stage' spaces for 'infra-politics ... the obtrusive realm of political struggle' (Scott 2005 [1997]: 311). The Cuban phrase heard very often in Tuta: '*Candidad de la calle, obscuridad de la casa*' (Openness from the [view of the] street, obscurity from the [view of the] house) probably means that one must be forthright about one's revolutionary conviction in public but that this may become muddled in private settings. Such 'hidden transcripts' (Scott 2005 [1997]: 314) play a significant role in social relations within private spaces in Tuta, which encompass both the physical household *and* unseen areas in town that may become important for securing household provisioning.

Food scarcities are a daily concern for Tutaños, and as already noted it is common to hear talk of getting foodstuffs on the street in revolutionary language, for example engaging in 'the fight for provisioning'. But when this talk of scarcities enters the private sphere, it often becomes what some habaneros have called a *tragicomedia* (tragicomedy; Tanuma 2007: 50) that seems to oppose rather than support the cultural politics of Cuban socialism. Yet consumer narratives heard in private spaces, such as Ricardo's joke about the 1959 'hurricane', are more complicated than they seem to outsiders. They are not just homogeneous acts of resistance against the national project, as suggested by James Scott's idea of house-hold 'infra-politics' (Scott 2005 [1997]: 311). Indeed, like consumer narratives of the *lucha* Cuban irony may be a way in which Tutaños identify with that very moral economy.

Sachiko Tanuma, another anthropologist of Cuba, has explicitly dealt with the topic of Cuban irony in her excellent paper dedicated to the subject (Tanuma 2007). She gauges the limits of Cuban humour by reading an ethnographic account[9] to Joanna, a Cuban sociologist. While Joanna first enjoyed the ethnographic detail of women engaging in 'the fight', stating: 'I really sympathize with the words of those women. I didn't know there were so many of them who thought a lot like me', she changed her mind when she heard the end of the ethnography, which states: 'Cuban women certainly live in a situation ... of profound failure to meet their expectations' (Tanuma 2007: 50, from Holgado Fernández 2000: 334). Responding to this conclusion the Cuban woman made 'a disgusted face, looked down, shook her head, and said in a determined voice, "*No me gusta*" (I don't like it)' (Tanuma 2007: 50). According to this account, though outsiders may associate language that targets the difficulties of the Revolution with unmitigated disapproval, insiders do not. Indeed, this example shows that such a connection may breed disapproval or moral qualms.

Sidney Mintz, the famous anthropologist of food in the Caribbean, apparently arrived at a similar conclusion. He criticized another ethnographic account which centred on the ineffectiveness of the Cuban state's distribution system instead of explaining how the state's failure to eradicate scarcities is understood by the people who actually have to live with them. When Tanuma met with Mintz in Japan, the latter related what he claimed to be a classic Cuban joke. It is very likely that this is the full version of Claudia's humorous comment about the three most important things in Cuba mentioned at the beginning of this section. 'What are the three successes of the Cuban Revolution?: Medicine, education and athletics. What are the three failures of the Cuban Revolution?: Breakfast, lunch and dinner!' Tanuma compares his use of irony to her interview with Joanna, the sociologist:

Of course, Mintz does know that food distribution in Cuba is far from perfect. He and Joanna share a sense of humor [*sic*] and make similarly ironical jokes. So why is it that they 'dislike' these ethnographic writings that are based on long-term fieldwork that also describe the current situation in Cuba in an ironical voice? ... [It is because] such writings ... often reduc[e] the irony of the formerly committed insider to the ironic narrative of the observing outsider. (Tanuma 2007: 51)

Unlike an outsider's account of Cuban realities, no matter how ironic, discursive portraits of failure such as the 'successes and failures' joke take place within *shared* moral spaces. As elsewhere, so in Cuba: 'The person who does the mocking lives in the same world as the mocked' (Tanuma 2007: 47). In Cuba, public rituals such as monthly syndicate meetings or speeches on commemorative days work to continually remind individuals not only of dominant norms, but also of collective values and morals. Mary Douglas contrasts such occasions with 'ritual pollution', or direct attacks on the value system which cause 'abomination ... which contradic[t] the basic categories of experience and in so doing threate[n] both the order of reason and the order of society' (Douglas 1975: 106). As Douglas asserts, 'ritual pollution' is not the same as private jokes; the latter are instead temporary suspensions of the moral order, or 'disturbance[s] in which the particular structuring of society becomes less relevant than another. ... *[T]he strength of [their] attack is entirely restricted by the consensus on which [they] depend for recognition*' (Douglas 1975: 106; my emphasis). While Cuban ironies may contain 'embryo[s] of another cultural order' (Sahlins 1976: 43), the very background information that enables such jokes to work for Cubans – though perhaps not for 'us' – is based on cultural and scalar politics of the nation state, about which Cubans are much more familiar. Non-Cubans may not find the 'successes and failures' joke particularly funny; however, such jokes are comical to Cubans (within Cuba) who are positioned similarly in relation to a nation state that has fallen short of collective ideals.

When considering Tutaños' ironic critiques of centralized distribution, therefore, one must not presuppose a strict opposition between the top-down values challenged in the joke and those of the individual who utters it. Using the term 'coevality' (borrowed from anthropologist Johannes Fabian 1983), Chris Gregory rejects strict dichotomies between subaltern and dominant representations (Gregory 1996: 204; 1997: 245). The idea of coevality is also relevant for Roger Lee (2006), who insists upon the multiple or coeval logics of economic practice, which exist both within and between individuals. Tutaño narratives of food consumption reflect such coeval moral, spatial and economic logics, some of which stem from individual or household pressures or desires, some from national values. To reduce Tutaño narratives of consumption to the exploiters/resistors binary reduces these multiple and shifting interfaces between citizens and the nation state to singular and static relations, excluding, among other things, the shared trajectories that have shaped Cuban identities. As Tanuma writes:

> While the jokes told by my friends may sound critical of and sarcastic toward the political leaders and the state, most of [my friends] had been faithful revolutionaries until the beginning of the Special Period. … Their shared sentiment then was not complete resentment toward the revolutionary ideas and institutions, but sad disillusion toward the present situation. What makes their … jocular comments ironic is *their half belief in and adoration of the revolutionary figures and … mythical stories that they are mocking.* (Tanuma 2007: 57; my emphasis)

Though people in Tuta may joke that 1959 was a 'hurricane' that wiped out all food in Cuba, the contradiction that results is not 'axiomatic' (Gregory 1997: 9): it is not a stark contrast between pro-revolutionary and anti-revolutionary, collectivism and individualism, the state and the household (or individual). As opposed to an axiomatic contradiction, the hurricane joke is more like what Chris Gregory calls a 'commonplace contradiction', one that reflects a particular balance between two or more viewpoints, a 'coexistence of rival cognitions' (Gregory 1997: 10–11). Cuban irony and other narratives of consumption referred to in this chapter and the next reflect multiple logics, some more individualized, some more collective. This insight will be important for understanding the relation between such cultural logics and ordinary economic practices, dealt with most thoroughly in chapters 5 and 6.

Negotiating individual desires with national norms: 'want' versus 'like'

Besides irony, another narrative that offers the outsider a glimpse of the moralities, materialities and spatialities of consumption is Tutaños' distinction between what one wants and what one likes, a contrast briefly outlined in this final section of the chapter. Perhaps more explicitly than Cuban irony, this simple example reveals that the relation between individual desires in Tuta and wider norms of the Cuban national moral economy is cultural.

In Tuta, it did not seem to be appropriate etiquette to express one's longing for commodities to others, perhaps because in Marxist thought this is associated with 'unnatural' material aspirations. Like the idea of a 'natural' market, which we have seen to be used in socialist as well as (neo)liberal contexts, the language of naturalism in Marxism is used to distinguish between good and bad forms of consumption (see chapter 4). While in the capitalist economy, the impersonal object – the commodity – is the foundation of individual–society relations, in Cuba's socialist economy, the commodity is seen to break down these very relations. In the communist economic model first advocated by Guevara, commodity value in the market sphere is entirely separate from labour value in the state sphere. Thus as we have seen in the previous chapter, in Ernesto Che Guevara's ideal society, money as well as unnecessary need should be eliminated altogether.

Despite actual materialities of commodification in present-day Cuba, at the scale of inter-personal relations, the commodity must be converted into a social form – the gift – before one is able to accept it as an object of consumption. This moral 'conversion' (Bohannan 1955, 1959; Bohannan and Bohannan 1968) transcends the boundary between the sphere of the market and that of 'mutuality' (Gudeman 2008), making what is usually unacceptable acceptable by drawing from overarching values in Cuban society. Thus while I never heard anyone – young or old alike – say 'I want' something, talk of what 'I like' was heard on a daily basis. Perhaps this is because the phrase 'I want' is unambiguously associated with mercenary desires while 'I like' *may* lead one to receive a commodity as a gift – enabling a moral conversion between market spaces and spaces for mutuality (for a parallel analysis of economic exchange in Cuba, see Tankha 2012).

Throughout the fieldwork period, I found that talk of 'what I like' was an admissible way to tell someone – especially a foreigner like myself – what 'I want'. For instance, it was not proper for Clara, my Cuban mother, to ask me directly for certain foods or drinks sold in hard currency, which she desired but could not afford. She had no problem, however, constantly reminding me that she 'likes' cola or pear juice, at least in private spaces (in chapter 5, I provide a contrasting account of foreigners' gifts in *public* spaces).

The moral conversion (between market and mutuality) also happened the other way around: when I took the role of receiver rather than giver. When once I said to a young woman, 'I like your blouse', I meant it as an off-hand compliment which is common (especially among women) in my own society. So I was surprised to hear the response: 'It is yours' rather than the 'thank you' one would expect in Euro-American contexts. I tested this 'compliment' on other young people and almost all responded in the same way, even embarrassing me a few times by giving me the item in question. If I 'liked' the clothes or adornments on one's body, this meant that the object should belong to me. To accept the compliment and then keep the object referred to seemed to mean that its owner valued it primarily as an impersonal commodity, rather than as an item with social value. As property disembedded from society, the blouse, necklace or hair band would then be seen as an unnecessary extravagance.

In this chapter, I have argued that scarcities of state goods and uneven access to commodities must be viewed through the lens of ordinary people in Cuba, and that we may better understand their ideas and practices of food provisioning if we pay attention to local uses of top-down representations like the idea of the national 'Fight'. Food consumption in Tuta is underpinned by coeval value systems that are variously individualistic and communistic according to the relations at play in any particular encounter. While we cannot romanticize the difficulties of living in a world without certain desirable (and necessary!)[10] items like beef, neither can we neglect the ways such extreme conditions are understood by the people who must live under them.

Cultural representations such as *lucha*, irony and expressions of desire allow Tutaños to shift between personal inclinations and wider normative modalities that stem from the moral economy of socialist Cuba. Despite variations in the ways individuals act in everyday settings, social institutions – often created by locals themselves, such as humour and etiquette – provide moral continuity and maintain the normative order. Official and localized values coexist, and people constantly switch from one to another (Gregory 1996: 204; 1997: 245). Between two ends of the spectrum – on the one hand following the rules, on the other breaking them – exists a grey area instituted by the people themselves. At the everyday level of household reproduction, Tutaños must be *bricoleurs*, creating new stuff out of given cultural material.

As we have seen in this chapter, so in the next: everyday moral economies of food consumption in Tuta reflect frustration as well as tolerance, conflict as well as conformity. Such bi-polar emotions exist not only between people, but within any individual person. At all scales, the Cuban Revolution is underpinned by a struggle between internal and external values and interests, a struggle that has characterized Cuba since the beginnings of its '*cubanía rebelde*' (ideology of dissent).[11]

Notes

1   Marston *et al.* (2005: 421).
2   Contrary to what many people believe, the United States sells a large number of commodities to Cuba, particularly basic foods and medicines. In 2000, former President Bill Clinton signed the Trade Sanctions Reform Act, which permitted the export of alimentary and medical commodities to Cuba, but prohibited US financing of such exports.
3   http://www.cubanet.org/CNews /y01/nov01 /27e1.htm (last accessed 17 May 2013).
4   See Gudeman and Rivera (1990: 28–9); Scheper-Hughes (1993: 158); Orlove (1997: 257); Pertierra (2007: 78, 2011: 109, 142–3).
5   One *quintal* is approximately 50 kg.
6   These points are also made in Wilson (2012b).
7   I thank Kenneth Olwig for pointing out that 'custom' derives from the Latin *moralis*. Though it refers to foundational legal principles, the idea of 'custom' (particularly in E.P. Thompson's sense) is entrenched in practice (Olwig, personal correspondence).
8   Roitman (2005: 21).
9   The ethnography, *¡No es Fácil! Mujeres Cubanas y la Crisis Revolucionaria (It's Not Easy! Cuban Women and the Crisis of the Revolution*; 2000), is by the Spanish anthropologist Isabel Holgado Fernández.

10   I claim, as do Cubans themselves, that beef is necessary for Cuban diets. This is particularly so given the prevalence of anaemia (particularly in the early to mid 1990s) and scarcities of adequate nutritional supplements in Cuba.
11   Kapcia (2000).

References

Álvarez, José. (2004a) Cuba's agricultural markets (Paper FE488). Institute of Food and Agricultural Sciences (UF/IFAS), University of Florida, http://edis.ifas.ufl.edu (last accessed 17 May 2013).

Álvarez, José. (2004b) Overview of Cuba's food rationing system (Paper FE482). Institute of Food and Agricultural Sciences (UF/IFAS), University of Florida, http://edis.ifas.ufl.edu (last accessed 17 May 2013).

Berg, Mette Louise. (2004) Tourism and the revolutionary new man: the specter of *jineterismo* in late 'Special Period' Cuba, *Focaal: European Journal of Anthropology* 43: 46–55.

Bohannan, Paul. (1955) Some principles of exchange and investment among the Tiv, *American Anthropologist* 57: 60–70.

Bohannan, Paul. (1959) The impact of money on an African subsistence economy, *Journal of Economic History* 19: 496–7.

Bohannan, Paul and Laura Bohannan. (1968) *Tiv Economy*. Evanston: Northwestern University Press.

Bohannan, Paul and George Dalton. (1962) Introduction. In *Markets in Africa*, edited by Paul Bohannan and George Dalton. Evanston: Northwestern University Press, pp. 1–26.

Boltanski, Luc and Laurent Thévenot. (2006) *On Justification: Economies of Worth* (transl.by Catherine Porter). Princeton: Princeton University Press.

Bourdieu, Pierre. (2005) *The Social Structures of the Economy* (transl. by Chris Turner). Cambridge: Polity Press.

Chayanov, Alexander V. (1966 [1925]) *Theory of the Peasant Economy*, edited by Daniel Thorner, Basile Kerblay and R.E.F. Smith. Manchester: Manchester University Press.

Douglas, Mary. (1975) *Implicit Meanings*. London: Routledge and Kegan Paul.

Fabian, Johannes. (1983) *Time and the Other: How Anthropology Makes its Object*. New York: Columbia University Press.

Fortes, Meyers. (1969) *Kinship and the Social Order. The Legacy of Lewis Henry Morgan*. Chicago: Aldine.

Gibson-Graham, J.K. (1996) *The End of Capitalism (as We Knew it): a Feminist Critique of Political Economy*. Minneapolis: University of Minnesota Press.

Gluckman, Max. (1965 [1955]). *The Judicial Process among the Barotse of Northern Rhodesia*. Manchester: Manchester University Press.

Granma, Official Organ of the Central Committee of the Cuban Communist Party. (2008) Consideraciones del Ministerio de la Agricultura sobre la producción y comercialización de productos agropecuarios. Article (no author) emailed to author, 4 June 2008, 1–4.

Gregory, Chris. (1996) Cowries and conquest: towards a subalternate quality theory of money, *Comparative Studies in Society and History* 38(2): 195–217.

Gregory, Chris. (1997) *Savage Money: the Anthropology and Politics of Commodity Exchange*. London: Harwood.

Gudeman, Stephen. (2008) *Economy's Tension: the Dialectics of Community and Market*. New York: Berghahn Books.

Gudeman, Stephen and Alberto Rivera. (1990) *Conversations in Colombia: the Domestic Economy in Life and Text*. Cambridge: Cambridge University Press.

Herskovits, Melville. (1962) Preface. In *Markets in Africa*, edited by Paul Bohannan and George Dalton. Chicago: Northwestern University Press, pp. vii–xvi.

Holgado Fernández, Isabel. (2000) *¡No es Facil! Mujeres Cubanas y la Crisis Revolucionaria*. Barcelona: Icaria.

Kapcia, Antoni. (2000) *Cuba: Island of Dreams*. Oxford: Berg.

Lee, Roger. (2006) The ordinary economy: Tangled up in values and geography, *Transactions of the Institute of British Geographers* (N.S.) 31: 413–32.

Marshall, Jeffery H. (1998) The political viability of free market experimentation in Cuba: evidence from *los mercados agropecuarios*, *World Development* 26(2): 277–88.

Marston, Sallie A., John Paul Jones III and Keith Woodward. (2005) Human geography without scale, *Transactions of the Institute of British Geographers* (N.S.) 30: 416–32.

Martí, Jose. (1975) *Inside the Monster: Writings on the United States and American Imperialism* (ed. and transl. by Philip S. Foner). New York: Monthly Review Press.

Orlove, Benjamin. (1997) Meat and strength: the moral economy of a Chilean food riot, *Cultural Anthropology* 12(2): 234–68.

Pertierra, Anna Cristina. (2007) Cuba: the struggle for consumption, doctoral thesis, University College London.

Pertierra, Anna Cristina. (2011) *Cuba: the Struggle for Consumption*. Coconut Creek: Caribbean Studies Press.

Richards, Audrey. (1932) *Hunger and Work in a Savage Tribe: a Functional Study of Nutrition among the Southern Bantu*. London: George Routledge and Sons.

Ritter, Archibald R.M. (1998) Entrepreneurship, microenterprise, and public policy in Cuba: promotion, containment, or asphyxiation?, *Journal of Interamerican Studies and World Affairs* 40(2): 63–94.

Roitman, Janet. (2005) *Fiscal Disobedience: an Anthropology of Economic Regulation in Central Africa*. Princeton: Princeton University Press.

Roland, L. Kaifa. (2006) Tourism and the *negrificatión* of Cuban identity, *Transforming Anthropology* 14(2): 151–62.

Sahlins, Marshall. (1976) *Culture and Practical Reason*. Chicago and London: University of Chicago Press.

Sayer, Andrew. (2003) (De)commodification, consumer culture and moral economy, *Environment and Planning D: Society and Space* 21: 341–57.

Scheper-Hughes, Nancy. (1993) *Death without Weeping: the Violence of Everyday Life in Brazil*. Berkeley: University of California Press.

Scott, James. (1998) *Seeing Like the State: How Certain Schemes to Improve the Human Condition Have Failed*. New Haven: Yale University Press.

Scott, James. (2005 [1997]) The infrapolitics of subordinate groups. In *The Postdevelopment Reader*, edited by Majid Rahnema and Victoria Bawtree. London: Zed, pp. 311–28.

Stein, Mary Beth. (1989) The politics of humor: the Berlin wall in jokes and graffiti, *Western Folklore* 48: 85–108.

Suárez Salazar, Luis. (2000) *El Siglo XXI: Posibilidades y Desafíos para la Revolución Cubana*. Havana: Editorial de Ciencias Sociales.

Tablada, Carlos. (1990 [1987]) *Che Guevara. Economics and Politics in the Transition to Socialism*. New York: Pathfinder.

Tankha, Mrinalini. (2012) Dismembering 'El Caballo': symbolic and economic meanings of multiple currency exchange in Cuba, Paper presented at the Annual Meetings of the American Anthropological Association, San Francisco, Nov.

Tanuma, Sachiko. (2007) Post-utopian irony: Cuban narratives during the 'Special Period' decade, *PoLAR: Political and Legal Anthropology Review* 30(1): 46–66.

Wilson, Marisa. (2012a) The moral economy of food provisioning in Cuba, *Food, Culture and Society* 15(2): 277–91.

Wilson, Marisa. (2012b) The moral geography of food in a dual economy: tourist versus domestic consumption in Cuba. *Caribbean Geographer* 17(1): 1–12.

Wright, Julia. (2008) *Sustainable Agriculture and Food Security in an Era of Oil Scarcity: Lessons from Cuba*. London: Earthscan.

Wolf, Eric. (2001) *Pathways of Power: Building an Anthropology of the Modern World*, with Sydel Silverman. Berkeley: University of California Press.

Yglesias, Jose. (1968) *In the Fist of the Revolution: Life in Castro's Cuba*. Middlesex: Penguin Books Ltd.

Chapter Four
Changing Landscapes of Care: Re-distributions and Reciprocities in the World of Tutaño Consumption

In this chapter, I continue to develop an understanding of Tutaño narratives of consumption, which comprise part of the normative background that shapes ordinary economic practices in Tuta. As in chapter 3, I argue that local understandings of scarcity and uneven access often draw from long-term values that have developed in Cuba over time, but that both connections and disconnections exist between Tutaño representations of consumer entitlements and socialist discourses of the nation state. I relate Tutaño narratives of consumption to recent shifts in Cuban policies, which partly counteract communistic relations between state and citizen. While Tutaños use expressions like 'we went hungry' to indicate their estranged relations with a state that seems to be failing to secure their needs, the new government has partly shifted its ideological emphasis from the communist value of universal needs (based on the tenet: 'from each according to his capacity, to each according to his needs') to the socialist value of universal work ('from each according to his capacity, to each according to his work').[1] According to this view, most recently espoused in Raul Castro's 2011 social and economic policies, Cubans who rely too much on the state without reciprocating with their labour or, increasingly, with their income from the market, are 'parasites' (Yaffe 2009: 269): unworthy of collective rewards since they do not contribute to the Revolution.

Still, the communist ideal of securing universal use values is far from obsolete in Cuban policies. As we shall see in the last part of this chapter, recent projections for economic progress outlined by Raul Castro's

Everyday Moral Economies: Food, Politics and Scale in Cuba, First Edition. Marisa Wilson.
© 2014 John Wiley & Sons, Ltd. Published 2014 by John Wiley & Sons, Ltd.

government in their economic reforms of 2010 and 2011 *are* consistent with Guevara's modernist view of economic transition, which emphasized *both* socialist *and* communist forms of value. While the former is maintained by the identification of Cubans as workers, the latter continues, if merely as a future ideal of just (re)distribution. Both forms of value – the one based on reciprocity between workers and the state, the other on the re-distribution of use values – rest on a scalar politics that requires the centralized production and re-distribution of productive resources (an *oikos*). And, as we shall see, this centralization and politicization of reciprocity and re-distribution is not unique to Cuba.

The chapter is divided into five brief parts. In the first, I offer a comparative view of reciprocity and re-distribution by drawing from anthropologists who have considered the centralization of food resources in other cultural and political contexts. In the second, I draw from this comparative perspective to argue that the Cuban government's responsibility to provide for workers is premised on political determinations of merit, which combine socialist ideas of reciprocity with hierarchical forms of re-distribution. The third and fourth parts of the chapter narrow the view of (re)distribution and reciprocal exchange to changing patterns of food and energy consumption in Tuta. Here, I provide ethnographic data to reveal how Tutaños adopt and change the government's discourse of merit through consumer narratives of hunger, nourishment and need. In the final part, I argue that hierarchical re-distribution and the reciprocal contract coexist and continue to support one another in recent Cuban policies, despite discursive shifts in landscapes of care from the state to the individual worker and his family. Like neoliberalizing discourses in other (post)socialist countries, we shall see how Raul Castro's policies for individual 'responsibilization' are buffered by 'inherited subjectivities of socialism' (Stenning *et al.* 2010: 35), particularly the identification of Cubans as, first and foremost, workers. Unlike the 'neoliberalization' of former socialist countries in Russia, eastern Europe and elsewhere, however, recent Cuban policies *also* maintain the communist value of centralized re-distribution, with financial regulations that resemble those implemented by Che Guevara at the very beginning of the Revolution.

Reciprocity and re-distribution: a comparative view

As suggested in the last chapter, in Cuba the paternalistic relationship between state as provider and citizens as workers links home consumption to the nation: Aristotle's *oekonomie* is bounded by the national territory. This relation is not just an official invention, for most Tutaños spoke of Cubans within the national territory as a single 'family' (see chapter 5). The 'domestication' (Creed 1998) of national discourses is not uncommon in centralized nation states, and there are more extreme examples elsewhere.[2]

For our purposes, the most important detail is the metaphor of the bounded familial unit which links hierarchical re-distribution with reciprocal exchange, an idea with origins in Aristotle's *Politics*.

Anthropologists such as Walter Neale have indicated that reciprocal relations and re-distributive hierarchies often coexist in societies; like individualism and collectivism, the articulation between the two rests on cultural foundations. Within Indian villages, for example, reciprocity between castes coexists with hierarchical re-distributions of grain; both are ordered by religious sanctions according to the specific moral economy of Indian society (Neale 1977 [1957]: 152). In this context, re-distribution itself may be regarded as reciprocity, as in the case of exchanges between a landlord and his dependent castes (Raheja 1988).[3]

Like anthropologists of India, classical anthropologists of Africa argued that the relationship between the person or group who gives and the person or group who receives food and other necessities reflects particular relations of power.[4] One emphasis of Audrey Richards's 1939 study, *Land, Labour and Diet in Northern Rhodesia* (now Zambia), was the structural significance of food distribution within African kin groups. As Richards argues, within the village kin group, the privilege of allocating cooked food is a way that sons-in-law gain authority within their wife's or wives' lineage. Special rights to receive enough cooked food to re-distribute are reserved for certain categories of people, such as the son-in-law. Legal standards and etiquette in this matrilineal society require that the mother-in-law has a large amount of cooked food sent to her son-in-law at the other side of the village, where he gains authority as distributor of cooked food amongst his own group of *affines* (relatives by marriage). The same rule is present when showing hospitality to visitors:

> The highest compliment [a man] can give his guest is to provide him with cooked food or beer, which he in his turn can distribute again. ... To give your elderly uncle a basket of porridge to eat behind closed doors means that you consider him as a chief who can eat alone and distribute food as he pleases. (Richards 1939: 129–36)

For Richards, the Bemba system of food distribution orders society on political, economic *and* moral levels. Givers of food have authority, but a taker gains authority by re-distributing what he has been given in turn. In accordance with Bemba rules of etiquette, anyone who takes food is morally obliged to reciprocate, perhaps by offering millet beer or porridge on another occasion (Richards 1939: 201–2).

In Cuba, where re-distribution and reciprocity are also valued over other economic forms such as the market, the person or entity that brings nourishment to his 'kin' is attributed a similar status of authority. And, as we have seen above, the taker in Cuban society – the worker – is morally obliged to reciprocate with his labour. As in other places (and times), ambiguities in

Cuba between re-distribution and reciprocity not only reproduce the centralized *oikos* but also maintain the political status quo. In most if not all places, moralizations of reciprocity often obscure ongoing hierarchies between givers and takers of food and other forms of welfare (Sahlins 2004 [1974]: 134). Thus in socialist as well as capitalist economies, reciprocal exchanges of work for state welfare overshadow the state's political authority to decide who gets what.

The rationale for the state re-distributive economy is that the centralized state can best:

> determine social needs and distribute in an equitable fashion according to objective needs. However, experience has shown that needs seldom present themselves objectively, but emerge only through a system of interpretation derived from the larger social and political discourse. ... The discourse may prioritize and legitimate certain needs, and ... it may construct criteria (such as ... moral-political criteria: past good or bad conduct ...) for assigning certain individuals and groups with special rights to the satisfaction of needs. (Yang 1989: 28)

The role of provider in places as diverse as India, former Northern Rhodesia, Cuba and even the United States (see below) is therefore a political role: it symbolizes the power to identify who merits provisions and who does not. Accordingly, in Cuban society the verb 'to nourish' (*alimentar*) symbolizes more than just feeding others. It also means the power to re-distribute resources to *specific* individuals – those identified as worthy – from the collective pot.

Merit and nourishment

The reciprocal relation between the worker's contribution and his or her due allows for unequal exchanges to persist disguised under the premise of merit. For instance, Article 7 of Cuba's new labour policy introduced in 2010, states that only the most 'suitable' (*idóneo*) state workers will remain in their jobs. The rest will be considered 'available' (*disponibles*) rather than unemployed, a terminology that reveals the continuous identification of *all* Cubans as workers for the Revolution. According to the policy, state workers who lose their jobs are paid 70% of their salary for up to 3 months, depending upon how many years of service they have served (Ministerio de Justicia 2010: 90–2). In rationalizing the state workforce, state managers (*jefes*) must avoid 'any manifestation of favouritism, like gender discrimination or other types [of discrimination]' (Ministerio de Justicia 2010: 90). Yet the decision of who to lay off does not ultimately rest with the *jefe*, but with a 'Committee of Experts', who are also in charge of determining who merits an augmented salary and who does not. Details of who makes up this committee are not provided in the policy, though it is made clear in a later policy document (Guidelines for the Social and Political Economy of the Party and the Revolution, April 2011, hereafter referred to as *Lineamientos* (Guidelines)) that the Party has the final word in implementing all such regulations (*Lineamientos* 2011: 38). The most definitive policy outlining current political and economic changes (Information about the Results of the Debate over the Guidelines for the Social and Political Economy of the Party and the Revolution, hereafter referred to as *Información*) was introduced in May 2011 after a public debate about the *Lineamientos*. *Información* mandates that salary increases will be granted only to workers 'whose labour has particular beneficial economic and social impacts [for the nation]' (*Información* 2011: 27). The criterion for bonuses is not further specified, but it is clear that the standards of value upon which it rests are continuous with the socialist moral economy, according to which workers who reciprocate with socially valuable labour are considered more worthy than those who do not.

Unlike the ideal meritocracy in capitalism, which rewards capitalists for increasing the general welfare of society as measured by ever-increasing levels of consumption and profits, in the Cuban economic model producers are compensated for increasing 'distributable resources' (Verdery 1996: 25). These are allocated accordingly, though often without any overall increase in consumption. Anthropologist of Romania, Katherine Verdery, provides a useful metaphor that illustrates the contrast between capitalist and socialist meritocracies:

> In capitalism, those who run lemonade stands endeavor to serve thirsty customers in ways that make a profit and out-compete other lemonade stand owners. In socialism, the point ... [is] not profit but the relationship between thirsty persons and the one with the lemonade – the Party center, which appropriated from producers the various ingredients (lemons, sugar, water) and then mixed the lemonade to reward them with, as it saw fit. ... [T]he transaction underscored the center's [*sic*] paternalistic superiority over its citizens – that is, its capacity to decide who g[ets] more lemonade and who g[ets] less. (Verdery 1996: 25)

Cuba's socialist vision, which rests upon the reciprocal relation between state and worker, obscures the fact that distinctions in the quality of labour – and therefore in hierarchical re-distribution – are 'attached to notions of personal evaluation' (Firth 1979: 185–7). Officials of the Cuban state are instilled with the authority to determine work allocation and salaries as well as distributive justice, and their idea of justice justifies distributing more social property to Party members and to those officially defined as advancing the aims of the Revolution (see chapter 5). Recent rationalizations of the

Cuban workforce must be viewed in the light of this idea of justice, which separates the 'worthy' who contribute to the Revolution from the 'parasites' who use its social property for their own benefit.

Furthermore, putting oneself in the position of provider (for the nation) without following the social norms this role entails is highly criticized by the revolutionary government. The Cuban state's role as provider may be employed by others, provided that s/he who assumes the position is able to identify those who merit provisions and those who do not, to distinguish who is a *luchador* (fighter) for the revolutionary cause and who is not. As in former Northern Rhodesia, one of the most valued actions in revolutionary Cuba is to supply food for others: the *campesinos* (peasants) who fed the rebels became symbols of the Revolution for this very reason. Present-day *campesinos* who donate to the Acopio are also recognized as 'combatants' who support the revolutionary cause.

By contrast, one of the biggest offences in revolutionary Cuba is to *alimentar* or 'nourish' those who are considered counter-revolutionary. The *campesinos* who chose to feed dissidents at the beginning of the Revolution were themselves deemed counter-revolutionary and all but eliminated by the second agrarian reform of 1962. 'Nourishing' those who do not fight for the Cuban version of justice strips the provider of authority and allies him or her with the enemy. Fidel Castro's 2007 outcry against George W. Bush involved a fierce critique of Bush's plan (announced in March 2007) to produce ethanol from corn and other crops, particularly in the 'developing' world. By 'nourishing' (*alimentando*; Castro used this term) cars for the 'First World', Bush was seen as supporting those who weigh personal wealth more than social justice and 'sentencing to death by hunger and thirst more than three billion people in the world' (Castro Ruz 2007: 1). According to Castro's critique, Bush wrongly took up the role of provider, re-directing resources from those worthy of nourishment – the world's hungry – to rich people in rich nations. For revolutionaries, nourishment symbolizes social justice.

'Hunger' in post-1990s Tuta: means testing for food and energy

The early 1990s' Special Period in Time of Peace, which officially ended by the late 1990s (though some argue it has not yet ended), was abnormally challenging according to many Tutaños' descriptions.[5] Prices for goods skyrocketed yet the average (official) salary remained at 300 pesos per month (a little less than $12 in 2011). Some Tutaños recalled that lard or vegetable oil, essentials for Cuban cooking, cost 100 pesos per litre during this period (in 2011, 1 L cost about 10–15 pesos). Another staple, rice, cost 45 pesos per pound (the 2011 price was 3.5 pesos per pound, about 7 pesos/kg). Accordingly, talk of dietary deficiencies during these years was common during fieldwork, especially when Tutaños were prompted on the subject:

'People lacked calcium and lost their fingernails … their hands showed signs'. And those who had lost too much weight were identified as one who 'is eating [electric] cable'.

At first glance, these accounts are consistent with findings from the Central Institute for Economic Planning of the United Nations Development Programme, which reported for the early 1990s a decrease in daily calorie and protein intake of 78% and 64%, respectively (quoted by Suárez Salazar 2000: 292). Longer-term studies, such as that conducted by the Statistics Division of the United Nations Food and Agriculture Organization, also indicate a higher percentage of undernourishment from 1993 to 1995 than in other countries in the region (22% compared to 12% for Latin America and the Caribbean as a whole; FAO-STAT c. 2006).

Despite their value, quantitative categories such as calorie and protein intake leave no room for *qualitative* accounts of scarcity and hunger, which are cultural. Indeed, as I argue for the case of energy rationing below, ideas such as 'going hungry' in Tuta refer not only to physiological pangs of hunger but also to more profound feelings of abandonment by the paternalistic state. As Michel de Certeau (1984 [1980]: xviii–xix) argues, consumption – even within defined domains, such as nutritional intake – is also a kind of production, a means by which social actors (re)create the physical and symbolic spaces in which they live. A few years earlier, Mary Douglas and Baron Isherwood made a similar point: 'Consumption activity is the joint production, with fellow consumers, of a universe of values' (Douglas and Isherwood 1978: 67).

As we have seen in the last chapter in regard to cultural ideas of scarcity, local conceptualizations of nourishment and its counterpart, hunger, are symbolic expressions of revolutionary values in Cuba. Like jokes about scarcity or local uses of *lucha*, ideas of nourishment and hunger in Tuta extend beyond the physiological or material realm. For Tutaños, hunger signifies a feeling of neglect, a break from socialist rules of 'familial' reciprocity. Pertierra has noted a similar attitude amongst Tivoliceros (residents of Tívoli, a neighbourhood of Santiago de Cuba): 'For consumers in the Tívoli and elsewhere in Cuba, to eat poorly is commonly understood to be a failing of socialist [re]distribution, and conversely to eat well is seen to be a triumph over the difficult obstacles faced in acquiring diverse and tasty food' (Pertierra 2007: 99–100).

Tutaños spoke of the Special Period as a time when 'we went hungry' in the context of not receiving one's fair share from the state. But the 'state idea' (Abrams 1988) has become largely detached from any localized entity that may be to blame. Its association with the impersonal form 'They' accords with its distant role as provider. Indeed, as other anthropologists of Cuba have suggested (e.g. Pertierra 2007: 100), Cubans talk about food as 'arriving' at the neighbourhood distribution centre (*la bodega*) as if the provisions had entered the ration store by the hand of a supra-human power.

Figure 4.1   The 'Year of the Energy Revolution' saw the proliferation of bicycles, including bicycle taxis. Source: author.

It is this very dependency on the larger-than-life state that the present government is calling into question, though, as I will soon reveal, still with a form of justification that emphasizes the kinds of reciprocal relations that emerged as part of the national moral economy even prior to 1959.[6]

The correlation between the idea of hunger, on the one hand, and a feeling of neglect, on the other, was evident not only when Tutaños spoke of food shortages, but also when they discussed energy scarcities and rations. In February of 2007, the ecological and political economic shift from gas to electricity was initiated under the 'Battle of Ideas' campaign, which began in 1999. The shift to more ecologically efficient home appliances was also an outcome of the 'Year of the Energy Revolution' (Yaffe 2009: 269; Figure 4.1). Electric appliances from China, such as low-energy refrigerators and pressure cookers, were 'distributed' (sold at marginal cost to Cubans with the new option of credit owed to the Central Bank) in March 2007. In a discussion with several Tutaño women from different households regarding the 2007 changeover from gas to electric domestic appliances in Cuba, one woman claimed that 'They' are working on a wind energy plant in a nearby town to replace the present system, which relies on petroleum imports (mostly from Venezuela), as well as domestic gas extraction from Cuban seas, but that 'nobody knows' (*nadie sabe*) when it will be finished. A woman in her mid-forties expressed her concern: 'For now we rely on the [electric] generators ..., which depend on gas to function. ... During this last storm the electric cables broke in front of the *panadería* [bread store]. In every storm the cables break in a different spot. These cables are older than I am!'

The shift away from gas and distribution of energy-efficient appliances was accompanied by an official means-testing scheme for electricity consumption to penalize all households using more than 300 kWh of electricity per month (compare this to the monthly electricity consumption of an average household in the United States: 11 280 kWh! See http://www.eia.gov/tools/faqs/faq.cfm?id=97&t=3). Similar to food rations,[7] which are increasingly allotted to families according to need, a minimum amount (100 kWh) of electricity is provided to each household at a subsidized monthly rate (1 peso/kWh).[8] Wealthier Cuban families with extra household items such as air conditioners must pay incrementally higher rates after the 100 kWh subsidized amount (2 pesos each from 101 to 150 kWh, 4 pesos each from 151 to 200 kWh, 6 pesos each from 201 to 250 kWh, and so on). In my humble family of three people (including myself) the monthly electricity bill was about 180 pesos, amounting to 140 kWh per month (100 at the 1-peso rate and 40 at the 2-peso rate). In line with the overall emphasis on the collective 'Fight', Cubans are encouraged to 'resist' by consuming as little energy as possible to conserve enough for collective re-distribution. Because wealthier citizens are expected to contribute more of their salaries for this purpose, electricity bills are generally more costly for high-income consumers. In this way energy rationing is comparable to food rationing, for the state is slowly eliminating the latter for higher earners who are encouraged to shop at CUC stores or higher-priced farmers' markets.

While the idea behind the 'Energy Revolution' is to reduce the country's reliance on fossil fuels while re-distributing wealth, the change from gas to electric has made life harder for poor people. Not only do most households in Tuta pay more for electricity than before (even with the subsidies), but women of these households connect energy shortages with difficulties preparing meals. Because of frequent power cuts during the fieldwork period, reliance on gas for cooking was unavoidable. When conversing about these changes, one woman complained that 'They' were not distributing enough gas for cooking. As there is not a wide variety of cold foods available in Tuta – bread and lunchmeat are much more expensive than tubers, rice and beans – cooking lunches and dinners is a daily task and most often the burden of women. Moreover, Cubans do not consider sandwiches and uncooked food a meal, and no matter how many snacks one has in a day, he or she must also have prepared food. Since Tutaños were used to receiving a quota of gas every 6 weeks from 'Them' (the government), a feeling of neglect by the national provider ensued when the household gas quota was reduced as part of the 'Battle of Ideas' campaign. The quota for gas that the state had distributed prior to the change, which easily lasted for a 6-week period, was, after March 2007, supposed to be kept as a reserve for the entire year. In most households I visited, the gas tank was already empty a month after the new quotas were distributed. Our neighbour expressed to me her concern about the issue: 'What if the electricity goes out and there is no gas in the house? Do we starve? When Hurricane Charley struck two years ago, Tuta had no electricity for a week. And the people without gas went hungry. *They* are letting us go hungry'. As these frustrated words illustrate, 'hungry' is used by Tutaños in a broad sense, indicating both physical hunger and a feeling that the state has abandoned them. Perhaps more than food scarcities, the recent shift from gas to electric has strained Tutaños' ability to 'resist', and many see a growing divide between their dedication to the state and what they receive in return.

Hunger and need

Just as the state as provider is partly losing its symbolic significance, so some people are increasingly at odds with official definitions of 'need'. Luis, in his forties, was disgusted by the decrease in food and gas rations: 'This is an imposition on our human right as consumers! If I want more, I cannot have it. Everyone *must* have the same amount!'

But nationalist views of need and re-distribution were also evident in local narratives of consumption. More than a few Tutaños with whom I discussed the issue saw means testing for food and energy as a positive move towards the kind of distribution that 'benefits the entire Cuban community'. For instance, Claudio (in his mid-forties) thought the 'Energetic Revolution' epitomized this collective ideal: 'Everyone will be equal! One will walk into another's house and say, "I have exactly what you have! I have a television of the same make, the same refrigerator, the same cooking appliances," and so on'.

While I have emphasized that many Tutaños uphold revolutionary values, even if just in the *calle* (street), I have also suggested that there are varying degrees to which individuals identify with official values. Often (but not always) the degree to which revolutionary values are upheld depends on a person's age. For most Tutaños of the younger generation, the contrast between 'before' (the 1970s and early to mid-1980s when food imports from the Soviet Bloc provided a wider variety of affordable goods) and 'after' (the post-Special Period era of food and energy scarcities) is the most vivid in their minds. But, as suggested in chapter 2, for many older Tutaños this disparity is not as significant as is the difference in Cuban life before and after the 'triumph' of the Revolution.

Like Claudio above, the older generation of revolutionaries and their followers, still arguably the primary voices in Cuban society, make a distinction between food (or energy) for all, or universal use value, and food (or energy) for a few. A common expression used in opposition to those who see the Special Period as the time when 'we went hungry' is: 'At least now [after 1959, rather than after 1989], no one goes to bed hungry'. An elderly man, Horacio, used the latter phrase on several occasions. He explained to me that the Revolution aimed to change people's pre-1959 consumption habits, teaching Cubans (most often on television and radio programmes) to 'act like people' rather than eating '*a lo loco*' (like a crazy person). In this sense, Horacio complimented the Revolution, indicating that there really was a change from striking inequalities before and equalities after: 'Before [1959], school children who attended classes with no breakfast or lunch had to sit alongside those who wore fancy clothes and ate sandwiches with a mountain of things inside'. Horacio applauded the government's early policy which established universal uniforms, prohibited children from taking their own meals and snacks to school and limited food consumption at school to what was available at the state-run cafeteria. The former president of the Union of Cuban Women for the Tuta municipality told me a similar story:

> Students who had money brought good snacks to school – bread with ham, chocolate milk – but most went without snacks or breakfast even. For a bit of energy, the poor kids would walk around sucking their *rollita de caña* [sticks of raw sugar cane], which they kept in their *cartúchos de papel* [paper sacks]. But these children were privileged to even go to school. Most could not go. ... Cubans today are still better off than [they were] before the Revolution: now everyone eats *something*, even if it is just *cualquier vulgaria*.[9]

For revolutionaries such as the woman quoted above, 'going hungry' is not the same as lacking a selection of desired foodstuffs; like other uses of the word (e.g. '"They" are letting us go hungry!'), 'hunger' signifies the inability to satisfy the needs of *society* rather than individual desires. The perspective of revolutionaries (at least) is consistent with Marx who distinguished 'natural' human needs from 'unnatural' material desires. This naturalist bias of Marxism places material necessity purely in the biological realm and ignores cultural definitions of need (Miller 1987: 46–8; Sahlins 2004 [1974]: 134–40).[10] Thus in Marx's view of consumption,

> productive activity comes before culture: use values – those desires and pleasures which spring from society – lose out to the objective means of their achievement. ... [T]he symbolic determination of needs ... is theoretically dissolved within the absolute objective action of their satisfaction. (Sahlins 2004 [1974]: 134–40)

The Marxist naturalization of need began early in the Revolution, as exemplified by official criticisms of fashionable clothing (especially US clothing), during the Vietnam War. José Yglesias's ethnographic-style account of his stay on the island in the early 1960s includes a fascinating description of the concept '*la enfermedad*' (the illness), as understood by locals in Mayarí, a town in eastern Cuba similar to Tuta in size and agricultural production (and also the hometown of Fidel Castro):

The first stage of the illness is non-political, then after longing for stylish clothing and obtaining some, and after pressure from ... [local law enforcement officials], at the last stage of the 'illness', they are counter-revolutionaries, 'they despise everything about the Revolution and are only waiting to reach that magical age of twenty-seven when they can apply to leave the country'. (Yglesias 1968: 202)

Yglesias interviewed young people at that time who do not seem so different from some of the youth in Havana city with whom I have interacted. It is important to emphasize, however, the significant difference in practices of and attitudes towards material consumption in Havana and Tuta. As noted in the introduction, many people in Havana are fully accustomed to the presence of foreigners and foreign commodities, as opposed to people in Tuta who rarely encounter foreigners unless they travel to Havana.[11]

In rural areas such as Tuta and Mayarí, conspicuous consumption leads others (officials *and* non-officials) to suspect that one is placing one's self-interest above the collective goals of the Revolution. Article 10 of the *Rules of the Cuban Communist Party* (1999) states that those who enter the Party must be characterized by an 'ideological firmness, modesty and simplicity, concretized in an austere form of living, without consumerist habits, in the energetic and intransigent defense of state property ...' (*Rules of the Cuban Communist Party* 1999: 53). And Che Guevara continues to represent the ideal member of society whose need for goods is purely utilitarian. School-age children in Cuba still must recite the daily adage: 'Pioneers for communism, we will be like Che!', and older Cubans enter the party under the premise that they will follow Guevara's example. Stories of Guevara's asceticism abound. One such narrative is of his wife Aleida, whom Che would admonish for using collective property to ameliorate her difficult work as a housewife:

[Aleida, who was] confronted daily, like many Cuban housewives, with the ordeal of long lines and shortages, sometimes used Che's official car escort, and [Che's] influence to fulfill the minimal requirements of subsistence. Che would send her to the market by bus, explaining irritably, 'No, Aleida, no, you know that the car belongs to the government, it is not mine, and you cannot take advantage of it. Take the bus, like everybody else'. (Ricardo Rojo 1968, cited by Castañeda 1998: 235–6)

I heard similar praises of asceticism – or, more likely, condemnations of self-interested behaviour – on the ground, especially from older revolutionaries. For instance, after the introduction of the dollar in 1993, many have seen an increase in the 'vices' of materialism. One elderly woman was outraged: 'The youth of today are distracted by material items. ... They all want to go to Florida. ... And when they finally make it [to Florida] they will have to work three jobs to keep up their habits of consumption!' In the context of a discussion about Serafín Alfonso, a young man who was allied with Castro prior to 1959 but then exiled as a counter-revolutionary in the first years of the Revolution, she said: 'When material restrictions start, the number of counter-revolutionaries rises. They are the people who do not have the spirit of sacrifice. They are opportunists!' In this account, as in the official value system, asceticism is associated with the revolutionary Fight for social justice.

While most Tutaños of the older generation faithfully defended the Revolution, at least in their words, we have seen that many of the younger generation regarded the situation in 2005–2007 and 2011 as contradictory. This is especially so in regard to a very particular material item: food. One young man in his early twenties, José, immediately introduced himself to me as a 'capitalist' when we met at a friend's house. The same person, however, used the 'work/enough food' cultural framework (introduced in chapter 2), to inquire about the quality of life in the United States. He did so by highlighting how well one eats 'over there': 'In the US, if one works, *at least* one eats, no? And I am sure that if one works hard, he eats *well*'. Instead of asking questions like 'what is the standard of life in the US?' or 'what is the cost of living over there?', as one would expect from a North American or British person, José correlated his idea of the 'good life' with food. Moreover, it was clear that José was not simply referring to eating for physiological purposes, but to eating *well*, a distinction based on preference and choice, which contradicts the Marxist 'natural' definition of needs.

Yet while José's comment countered this communistic ideal of universal use value, his critique also incorporates key symbolic ideas embedded in the history of revolutionary Cuba, particularly the reciprocal relationship between work value and alimentary remuneration. Indeed, though he extended his idea of 'need' to include a wider variety of preferred foods, José's reference to the 'system of labours' still corresponds to a 'system of needs' (Sahlins 2004 [1974]: 151), a relation that is not too far from the ideal socialist contract.

Continuities and change in the social contract

> The economic battle constitutes today, more than ever, the principal task and central ideological work of our party officials, because the sustainability and preservation of our social system depends on it. (*Lineamientos* 2011: 1; my translation)[12]

The socialist idea of reciprocity is justified by the persistent claim, first made by Ernesto Che Guevara, that national progress in both productive output and 'consciousness' will eliminate the need for rations and other subsidies. This official line is made manifest in recent policies, which 'guarantee the continuity and irreversibility of Socialism [*sic*], economic development of the country and the elevation of the quality of life of the population, along with the necessary formation of moral and political values of our citizens' (*Lineamientos* 2011: 5). Recent policies state that rations and subsidies will 'gradually' be eliminated as the 'duties and revolutionary sensibilities … determined by everyday behaviour' (*Lineamientos* 2011: 9) allow for an increase in the domestic production of food and other necessities.

This teleology aligns with what Raul Castro's government now scathingly calls egalitarianism: the 'equal' distribution of state resources to those who earn more than others through market sales or other sources (such as remittances from abroad), without the 'consciousness' to reciprocate enough in return. In other words, citizens who do not support the Revolution through work or taxes are not considered worthy of public goods such as food rations. Raul's reference to some Cubans as 'parasites' coincides with this hierarchy of merit: Cubans who work hard for social property are opposed to Cubans who earn from market networks. The latter are generally suspected of taking from collective property rather than adding to it through their work effort: 'the concentration of property will not be permitted' (*Lineamientos* 2011: 11). In fact, though recent policies reject 'egalitarianism' of distribution, they *do* propose an 'equality of tax duties', such that if one earns more, they pay more (*Información* 2011: 11–12). This kind of thinking underpins the way food and energy rationing are instituted in Cuba, as we have seen. Similar to the relation between re-distribution and reciprocity in other places and times, moralities of reciprocity in Cuba coexist with re-distributive hierarchies of merit, which are, in turn, based on how much one gives back to society through productive or monetary contributions. Cuban citizens who cannot work (and who are not *disponibles*) and who cannot get help from family or friends are considered worthy of state re-distribution (see *Información* 2011: 28), but the majority who can work must do so under the socialist contract.

It is clear that present policies in Cuba resemble welfarist reforms elsewhere, such as those in the United States in 1996 when president Bill Clinton introduced the Personal Responsibility Work Opportunity and Reconciliation Act. As in other countries (e.g. the UK) where similar legislation has been enacted, Clinton's policy shifted emphasis from welfare to 'workfare', a shift that was also premised on the value of working to 'give back' to the national 'community'. Perhaps the most ironic resemblance between Raul's policies and others in the capitalist world is his use of neoliberal discourses of dependency, which were also adopted in post-socialist countries of eastern Europe (Stenning *et al.* 2010: 1). Raul's reference to Cubans as 'open-mouthed hatchlings' (Dilla 2011: 1) recall these discursive shifts in the post-socialist world, but also those that occurred in the United States and the UK where the 'equality of welfare' discourse of the 1930s–1980s became an 'equality of opportunity' discourse in the 1990s (Young 2011: 180). Like such rhetorical shifts, Raul Castro's emphasis on equality of opportunity epitomizes what Iris Marion Young calls the 'blame' model of economic or legal discourse (Young 2011: 11), which targets the 'socially deviant' (Young 2011: 11) who receive more than they give.

In spite of these profound resemblances between recent Cuban policies and those in more neoliberalized contexts, Cuban policies reinforce rather than undermine the cultural and material unity of Cuba's socialist (and communist) project. Indeed, though neoliberal transitologists argue that recent initiatives in Cuba such as rewarding good workers with bonuses illustrate an 'incipient' (Feinberg 2011: 97) shift towards global capitalist relations, such policies (initiated in 2008) continue to be justified in terms of use values for the national Revolution rather than exchange values for the maximization of individual utilities. As Helen Yaffe argues:

> Capped or not, bonus payments continue to be awarded for over completion of the national plan in the production of physical goods and services – that is, in terms of use values, not in terms of exchange values. The plan is discussed in workers' assemblies and formulated according to political priorities, not market forces. … What lies behind the new wage incentives is not a return to capitalism but an effort to reduce Cuba's vulnerability to the global crisis resulting from a rise in food and fuel prices. … Far from representing a profound shift in the structure of Cuban society, the 2008 measures were taken to preserve existing state welfare provisions and strengthen the socialist economy. (Yaffe 2009: 271)

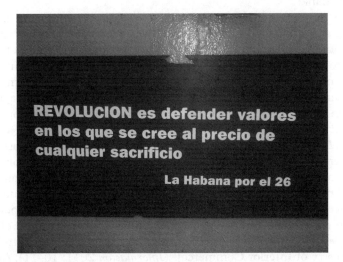

Figure 4.2   'Revolution is defending the values that we believe in at the price of any sacrifice'. Source: author (my translation).

Thus while political discourse in Cuba has shifted away from an equality of needs and towards an equality of opportunity, this is not an equality of opportunity in the *individualized* market, but an equality of opportunity in *collective* work/rewards.

Current policy shifts in Cuba contradict more communistic understandings of the state as provider for its citizens, a contradiction that Tutaños themselves lament. However, new discourses introduced in present Cuban policy are still continuous with the kinds of socialist and communist values first promulgated by Ernesto Che Guevara during the Great Debate of 1963/5. Yaffe (2009: 267–9) illustrates this continuity in recent Cuban financial regulations. Comparing current Cuban finance to that in 1961 when Guevara authorized the complete replacement of all Cuban banknotes with revolutionary pesos, Yaffe claims that 2003 saw a *re*centralization of the Cuban financial sector, such that all hard currency flows into the country were eventually converted into CUCs (Yaffe 2009: 267–9). And, as in the early 1960s, in August 2004 Cubans were given just 3 weeks to exchange their US dollars for CUCs, a restriction that Yaffe (2009: 268) attributes to a similar interest in confiscating funds that may have otherwise been used for counter-revolutionary activities. Moreover, in January 2005 all state enterprises (including Cuban investments in joint ventures) had to transfer their CUCs to the Central Bank, and in August 2007 all their profits had to be re-directed to the same institution (Yaffe 2009: 268). The centralized control of profits and money coming from abroad thus continues to be a primary policy for national development and 'just' (re)distribution, just as it was when Guevara was president of the National (now Central) Bank.

Profits of the increasing number of private entrepreneurs are also checked and redirected to the centre: '[we] need to establish a tributary culture and social responsibility ... to develop the civic value of contributing to social expenditures' (*Información* 2011: 11–12). Recent Cuban policies not only authorize the Central Bank of Cuba to determine the 'functioning' – or ultimate aims – of all economic entities in Cuba, including 'state enterprises, cooperatives, the private sector and the population' (Ministerio de Justicia 2010: 113), but also prioritize the plan over the market (*Lineamientos* 2011: 5). The importance of the plan is most apparent in centralized pricing policies for the domestic market in public goods deemed necessary. In line with Guevara's policies for tempering (and eventually eliminating) the market law of value as a measure for remuneration and prices, the Cuban government aims to 'establish a correspondence between the amount of money in circulation and the circulation of goods in the *Mercado Minorista* [domestic market in national pesos, controlled by the Ministry of Interior Commerce]' (*Información* 2011: 10–11). And like Guevara's pricing policies, the costs of goods and services 'that it is of economic and social interest to regulate' will be controlled centrally; the 'rest will be decentralized' (*Información* 2011: 12). The material circuit of public goods in the domestic sphere is thus firmly disconnected from the material circuit of commodities that enter into and out of the global market economy.[13]

Perhaps most importantly for Tutaños, recent policies project the 'gradual' unification of Cuba's two currencies. Presumably this will be achieved by eliminating the national peso altogether and keeping CUCs, though because the two are pegged to the US dollar like pre-revolutionary pesos, it will be interesting to see whether and how Cuba manages monetary unification without further exacerbating inequalities. One way of ensuring 'just' transition is through price controls, and indeed 2011 policies indicate that before monetary unification, and even before rations and other subsidies can be significantly reduced, Cuba needs to control prices. This means that prices of goods sold in national pesos (in the *Mercado Minorista*) will be stabilized and probably replaced by CUCs, but also that the prices of goods from abroad (distributed through the *Mercado Mayorista*, or Ministry of Exterior Commerce) will be controlled, particularly 'those associated with the basic necessities of the population' (*Información* 2011: 13). Again, the two material circuits of value remain decisively separated.

As in the early Revolution, for the present government, the only way to achieve these results is through universal work effort and an increase in revolutionary consciousness:

> The objective of creating a Cuban economic model is to guarantee the continuity and irreversibility of Socialism, economic development of the country and the elevation of the quality of life of the population, along with the necessary formation of moral and political values of our citizens. ... The policies define an economic system that prioritizes a continuation of socialist property of all the people as a fundamental aspect of production, which accords with the socialist distribution principle: 'from each according to his capacity, to each according to his work'. (*Lineamientos* 2011: 5; my translation)

A Soviet guide for Cubans entitled *What is Property?* outlines this Rousseauian exchange relationship between state and citizens, which continues to shape Cuban policies despite recent discursive emphasis on 'cultures of dependence' (Stenning *et al.* 2010: 1):

> The participation of members of socialist society in distribution and, respectively, consumption can be obtained only through work. Every single worker receives from society ... a quantity exactly equivalent to what he has given. ... Work that individuals engage in is both theirs and society's ... As a result, distribution according to work is an objective necessity, an economic law of socialism. (Suvorova and Románov 1989; my translation)

Figure 4.3   A queue outside a neighbourhood rationing centre. Source: author.

The moral obligation of the worker to produce a surplus for social needs is not unique to Cuba as a socialist country (Humphrey 1998: 76; Hivon 1998: 39) though this was a legal requirement that Fidel Castro's rebel army established even before the 'triumph' of the Revolution, and certainly before the official adoption of the communist line in 1961. Continuous with the re-distributive aims of Cuba's first revolutionary army, Castro and Guevara sought to institutionalize the production and allocation of land and food in the Sierra Maestra. Like later agrarian reforms (see chapter 6), Law No. 3 of 1957 obliged those who owned plots of land in the Sierra to produce for the revolutionary cause; those who refused to do so risked confiscation of their lands (Valdés 2003: 1).

After the 'triumph' of the Revolution, exchanges of centrally allocated food for individualized work efforts 'jumped scale' from the eastern coastal range to the entire nation state, institutionalized through the Acopio, rationing and state subsidies for basic necessities (Figure 4.3). The reciprocal relation between the revolutionary government and its citizens also 'jumped scale' in 1959, when it was made to fit the scale of the nation. Re-distributive policies, and the norms underpinning them, contributed to a scalar shift in Tutaño ideas of reciprocity, from working to earn 'enough food' for the household to working for the Revolution to satisfy the needs of the national *oikos*.

The socialist value of reciprocity – work for social entitlements – has overshadowed the revolutionary economic model in Cuba at least since Law No. 3 was enacted in the Sierra Maestra in 1957. But the idea of need as constructed by the Revolution is often at odds with that of a new generation of Cubans who must balance 'unnatural' desires with long-established political norms. Like the example of Cuban irony and other consumer narratives of the 'Fight', the way Tutaños express such contradictions between the normative and the actual is tied to specific cultural representations that have evolved in Cuba over time.

Notes

1   I thank my DPhil examiners, Paul Dresch (Oxford) and Martin Holbraad (UCL), for pointing out the discrepancy between these two versions of the communistic slogan.

2   One example is provided by James Watson in his account of 'eating from the big pot' at specially instituted mess halls in socialist China during the Great Leap Forward (1958–1961). According to Watson, official attempts to merge all families into one national family led to 'policy-induced famine' (Watson 2011: 37): 'New style families' replaced the unity of the kin group, and the preparation of meals within the home was all but eliminated as farmers were encouraged to donate their iron cookware and furnaces to collective eating establishments. Watson argues that universal food distribution in mess halls was a way to inculcate the idea of the nation as a single family or entity'.

3   I thank Bob Parkin of the Institute of Social and Cultural Anthropology (Oxford) for pointing me to this source.

4   That is, Richards (1932, 1939); Fortes and Fortes (1936); Firth (1959).

5   See appendix 2 for the per capita intake of calories, proteins, and fat for the 1980–1999 period.

6   As early as 1957, Fidel Castro and Che Guevara enacted their first revolutionary policy of re-distribution, mirroring 19th-century attempts by the rebel army to reorganize the countryside, which were thwarted by General Weyler's reconcentration policy (I refer to this policy at the end of the chapter).

7   Approximate monthly rations per person in 2013 are provided in appendix 4.

8   The following figures are based on interviews during the 2011 period of research and may have since changed.

9   '*Cualquier vulgaria*' literally means 'any vulgarity'. It is a common expression in Tuta used in reference to a small amount of anything, usually food.

10 I would argue, however, that the Marxist view of need is itself cultural in this context.

11 A full analysis comparing attitudes to consumption and materiality in rural versus urban Cuba is certainly wanting.

12 Translations of all citations from recent policies are my own.

13 See appendix 3 for a flow chart of institutional levels for national food provisioning.

References

Abrams, Philip. (1988) Notes on the difficulty of studying the state, *Journal of Historical Sociology* 1(1): 58–89.

Casteñeda, Jorge G. (1998) *Compañero: the Life and Death of Che Guevara*. New York: Vintage Books.

Creed, Gerald. (1998) *Domesticating Revolution: From Socialist Reform to Ambivalent Transition in a Bulgarian Village*. University Park: Penn State University Press.

Cuban Communist Party. (1999) *Rules of the Cuban Communist Party*. Havana: Editorial Política.

De Certeau, Michel. (1984 [1980]) *The Practice of Everyday Life*. Berkeley: University of California Press.

Dilla, Alfonso Haroldo. (2011) Cuban government deals with 'unusual situations', *Havana Times*, 21 Sept.

Douglas, Mary and Baron Isherwood. (1978) *The World of Goods: Towards an Anthropology of Consumption*. London: Allen Lane.

FAO-STAT (c. 2006). Food deprivation trends: mid-term review of progress towards the World Food Summit target. Statistics Division Working Paper Series. http://www.fao.org/docrep/013/am064e/am064e00.pdf (last accessed 16 May 2013).

Feinberg, Richard E. (2011) *Reaching Out: Cuba's New Economy and the International Response*. Washington DC: Latin American Initiative at Brookings.

Firth, Raymond. (1959) *Social Change in Tikopia: Re-study of a Polynesian Community after a Generation*. London: George Allen and Unwin.

Firth, Raymond. (1979) Work and value: reflections on ideas of Marx. In *Social Anthropology of Work*, edited by Sandra Wallman. New York: Academic Press.

Fortes, Meyers and S.L. Fortes. (1936) Food in the domestic economy of the Tallensi, *Africa* 9: 237–76.

Hivon, Myriam. (1998) The bullied farmer: social pressure as a survival strategy? In *Surviving Post-Socialism: Local Strategies and Regional Responses in Eastern Europe and the Former Soviet Union*, edited by Sue Bridger and Francis Pine. London: Routledge.

Humphrey, Caroline. (1998) *Marx Went Away but Karl Stayed Behind* (updated edition of *Karl Marx Collective: Economy, Society and Religion in a Siberian Collective Farm*, 1983). Ann Arbor: University of Michigan Press.

Información. (2011) *Información sobre el Resultado del Debate de los Lineamientos de la Política Económica y Social del Partido y la Revolución*. VI Congreso del Partido Comunista de Cuba. Havana, May.

Lineamientos. (2011) *Lineamientos de la Política Económica y Social del Partido y la Revolución*. VI Congreso del Partido Comunista de Cuba. Havana, 18 Apr.

Miller, Daniel. (1987) *Material Culture and Mass Consumption*. Oxford: Basil Blackwell.

Ministerio de Justicia. (2010) *Gaceta Oficial de la Republica de Cuba*, special edition. Havana, 9 Oct.

Neale, Walter. (1977 [1957]) Reciprocity and redistribution in the Indian village. In *Peasant Livelihood: Studies in Economic Anthropology and Cultural Ecology*, edited by Rhoda Halperin and James Dow. New York: St Martin's Press.

Pertierra, Anna Cristina. (2007) *Cuba: the struggle for consumption*, doctoral thesis, University College London.

Raheja, Gloria Goodwin. (1988) *The Poison in the Gift: Ritual, Prestation and the Dominant Caste in a North Indian Village*. Chicago: University of Chicago Press.

Richards, Audrey. (1932) *Hunger and Work in a Savage Tribe: a Functional Study of Nutrition among the Southern Bantu*. London: George Routledge and Sons.

Richards, Audrey. (1939) *Land, Labour and Diet in Northern Rhodesia: an Economic Study of the Bemba Tribe*. London: Oxford University Press.

Sahlins, Marshall. (2004 [1974]) *Stone Age Economics*. London: Routledge.

Stenning, Alison, Adrian Smith, Alena Rochovská and Dariusz Świątek. (2010) *Domesticating Neo-liberalism: Spaces of Economic Practice and Social Reproduction in Post-Socialist Societies*. Oxford: Wiley Blackwell.

Suárez Salazar, Luis. (2000) *El Siglo XXI: Posibilidades y Desafíos para la Revolución Cubana*. Havana: Editorial de Ciencias Sociales.

Suvorova, M. and B. Románov. (1989) *¿Que es la propiedad?* Moscow: Editorial Progreso.

Valdés, Orlando. (2003) *Historia de la Reforma Agraria en Cuba*. Havana: Editorial Ciencias.

Verdery, Katherine. (1996) *What Was Socialism, and What Comes Next?* Princeton: Princeton University Press.

Watson, James. (2011) Feeding the Revolution: public mess halls and coercive commensality in Maoist China. In *Governance of Life in Chinese Moral Experience: the Quest for an Adequate Life*, edited by Arthur Kleinman and Everett Zhang. London: Routledge, pp. 33–46.

Yaffe, Helen. (2009) *Che Guevara: the Economics of Revolution*. London: Palgrave Macmillan.

Yang, Mayfair Mei-Hui. (1989) The gift economy and state power in China, *Comparative Studies in Society and History* 31(1): 25–54.

Yglesias, Jose. (1968) *In the Fist of the Revolution: Life in Castro's Cuba*. Middlesex: Penguin Books Ltd.

Young, Iris Marion. (2011) *Responsibility for Justice*. Oxford: Oxford University Press.

Chapter Five
Localizing the Leviathan:
Hierarchies and Exchanges that Connect
State, Market and Civil Society

In chapters 3 and 4, I explained how representations that stem from the national moral economy are used on an everyday level to justify or contest present changes in the Tutaño world of *consumption*. This chapter will reveal how similar tropes of the national Leviathan, such as the idea of someone with or without revolutionary 'culture', work as levelling mechanisms to carve out spaces for 'just' *exchanges* and 'acceptable' inequalities in everyday life. Similarly, in chapter 6, I outline how the national moral economy shapes the way food is *produced* in Cuba.

In the moral community established by the Cuban Revolution (see Figure 5.1 for the official definition of 'Revolution'), private interest is checked by public morality and individuals are unified by a collective, historical drive towards a 'just' future. The idea of social justice in Cuba is evident from the following excerpt from a Soviet guide for the Cuban economy (translated into Spanish in 1965), which is still used for ideological education in Tutaño schools:

> In socialist society, the personal motivations of man are not opposed to the inter-
> ests of society, instead they are in agreement … Personal interests, moreover, are
> not suppressed in a society of this type; on the contrary, they reach their most
> complete expression and development, because they are in total agreement with
> the interests of society. … It is this harmony between personal and social interests,
> and their very unity, that disprove theories which claim that social influence can
> only be manifested as a suppression of the personal aspirations of men. (Smirnov
> 1965: 8–9; my translation)

Everyday Moral Economies: Food, Politics and Scale in Cuba, First Edition. Marisa Wilson.
© 2014 John Wiley & Sons, Ltd. Published 2014 by John Wiley & Sons, Ltd.

Revolución es: sentido del momento histórico; es cambiar todo lo que debe ser cambiado; es igualdad y libertad plenas; es ser tratado y tratar a los demás como seres humanos; es emanciparnos por nosotros mismos y con nuestros propios esfuerzos; es desafiar poderosas fuerzas dominantes dentro y fuera del ámbito social y nacional; es defender valores en los que se cree al precio de cualquier sacrificio; es modestia, desinterés, altruismo, solidaridad y heroísmo; es luchar con audacia, inteligencia y realismo; es no mentir jamás ni violar principios éticos; es convicción profunda de que no existe fuerza en el mundo capaz de aplastar la fuerza de la verdad y las ideas.

Revolución es unidad, es independencia, es luchar por nuestros sueños de justicia para Cuba y para el mundo, que es la base de nuestro patriotismo, nuestro socialismo y nuestro internacionalismo.

Concepto de Revolución

CIMEX

Figure 5.1 'Revolution is: A sense of duty to the historical moment; it is to change everything that must be changed; it is absolute equality and freedom; it is to be treated and to treat others as human beings; it is to emancipate ourselves by our very selves and with our own efforts; it is defiance against powerful dominant forces inside and outside our social and national environment; it is to defend the values that we believe in at any price and with any sacrifice; it is modesty, unselfishness, altruism, solidarity and heroism; it is to fight with audacity, intelligence and realism; it is to never lie nor violate ethical principles; it is the profound conviction that there exists no force in the world capable of crushing the force of the truth of ideas. Revolution is unity, it is independence, it is fighting for our dreams of justice for Cuba and for the world. This is the basis of our patriotism, our socialism and our internationalism' (my translation). Source: author.

In the communist economic model, the 'part of the person ... involved in the economic process' (Bloch 1989: 173) is inseparable from the value of the collective as a whole. In Louis Dumont's terms, the interests of both individuals and households are 'encompassed' (Dumont 1980 [1966]) by the scalar project of national economy.

In this ideological context, the collective is the national 'family'. Indeed, as Antoni Kapcia (2000: 240) argues, since the Special Period in Cuba there have been increasing references to the 'Cuban family' in official discourse. Given recent pressures on the biological or extended family (*familiares*), however, there are times and spaces when food that is 'good to eat' for one's *familiares* may not be 'good to think' for the national family (Harris 1986). Moral rules and 'the eyes' (*los ojos*) of officialdom restrict local provisioning of otherwise agreeable foods, such as pizza from a *paladar*. In ordinary economic life, however, there are times and spaces when and where these rules are broken.

Moral and legal rules set at the national scale often coincide with more localized social relations of value, reproducing economic geographies tied to the national value system. In the local moral economy, for instance, community gossip and other levelling mechanisms work to inhibit self-interested behaviour (see Gluckman 1963; Foster 1965; Paine 1967; Wilson 1973). For instance, Tutaños with larger homes or those engaging in expensive household construction complained to me about neighbours or others who 'think we are rich, but we are not'. Such talk becomes 'dangerous' (*peligroso*) when overheard by neighbourhood informants who pass this information on to upper levels.

Local rules that employ national values not only determine the boundaries of illegally licit[1] exchanges in Tuta between people and households and public and private spaces, but often perpetuate official hierarchies of merit, as introduced in chapter 4. The boundaries and possibilities of 'ordinary' economic geographies are thus limited by a national value system that provides moral justifications for economic inequalities entrenched in state, market and civil society. 'Lay' moralities (Sayer 2011) shape the cultural parameters of acceptable exchanges between ordinary people and privileged persons with political capital, *particulares* (self-employed entrepreneurs) and their clients, or between Tutaños and tourists. In this chapter, I take each of these relations of exchange in turn, but first continue along the lines of the previous chapter by explaining more thoroughly the institutional and ideological hierarchies that provide the cultural 'background' for such relations. The first half of the chapter deals with fundamental institutional and ideological factors that enable Cuba's scalar politics to work at the local level. In the first subsection, I describe institutions and ideologies that beset communist hierarchies. The second subsection reveals how public institutions enter into private spaces, such as the household. The final subsection of part 1 shows how institutional and ideological hierarchies are justified in official terms, particularly through the idea of revolutionary 'culture'. In the second half of the chapter, which is divided into four subsections, I use ethnographic data to uncover how top-down hierarchical structures become localized through everyday processes of valuation. I describe 'appropriate' exchange relations for *jefes* (state manager) (subsection 1) and then address cultural rules that moralize how *particulares* practice economy (subsection 2). The final two subsections deal with exchanges between the most (*familiares*; extended 'kin') and least (*jinetero/as*; street hustlers or prostitutes) acceptable exchange partners in Tuta.

Figure 5.2   Cuban youths sitting along the waterfront (*malecón*) in Havana. Source: author.

Institutions and ideologies of the national moral economy

Democratic centralism and scalar hierarchies

As in Dumont's (1980 [1966]) India, individuals and families in Cuba are encompassed by the hierarchical whole of Cuban communist institutions, in political and moral as well as practical terms. From the age of 4 or 5 years, each Cuban is systematically connected to larger collective levels, even if he or she is not a member of the Party. All Cubans are 'oriented' by official institutions such as the *Partido Comunista Cubano* (Cuban Communist Party, PCC) and by so-called unofficial institutions such as mass organizations. State workers are subdivided into types by the *Comité de los Trabajadores Cubanos* (Committee of Cuban Workers, CTC). Other categories of persons, such as *campesinos* (small farmers) and *particulares* are 'oriented' by cooperatives and associated mass organizations, such as the CTC. Plans for the growing number of non-state workers do not exempt the requirement to be 'integrated' by such hierarchical institutions.

The type of syndicate to which a worker is required to belong accords with the type of production they engage in; for example, the agricultural section of the CTC consists of a sugar cane syndicate, a tobacco syndicate, a mixed crops syndicate, a floral and ornamental plants syndicate, and perhaps others. During syndicate meetings some encouragement for production is offered via 'emulation', whereby exceptional workers are recognized and rewarded with moral incentives (e.g. certificates of participation or certificates of merit for producing over the plan) and/or material incentives (e.g. trips to tourist areas such as Varadero Beach). Other attempts to raise the political, social, cultural and economic goals of workers are exemplified by workshops in Marxism-Leninism and by symbolic ceremonies and speeches commemorating special days, such as the 'Day of the Farmers'(*Día de los Campesinos*).[2]

According to Peter Roman (2003), democratic centralism is a fundamental concept on which the Cuban economy is modelled (Roman 2003: 94), though references to it were eliminated in the latest (1992) constitution. The idea stems from Karl Marx, and before him, Jean-Jacques Rousseau, both of whom criticized the separation in classical economy between civil and political liberty, on the one hand, and social liberty, on the other (Roman 2003: 10–16). According to both theorists, this separation was based on the idea of private property, a key concept in classical economics. In the model of democratic centralism, private property is eliminated and becomes collective property, just as private interests no longer guide public decisions. In the Marxist version of democratic centralism, all workers are seen to have equal access to collective property created at work and changes to one's 'work unit' (workplace) cannot occur except through workers' mandates.

Workers (and now, other categories, such as *particulares* and *campesinos*) must be organized by the Party. Article 2 of the *Rules of the Cuban Communist Party* (Cuban Communist Party 1999), a primary guide for militants, states that all work units must have one (and only one) 'centralized direction' of the Communist Party: a 'work nucleus' that consists of at least three militants of the Party (Cuban Communist Party 1999: 4; as opposed to a work unit which is simply a workplace, as noted above). The work nucleus for adult militants is equivalent to the 'base organization' for members of the *Unión de Jóvenes Comunistas* (Young Communist Union, UJC). Each represents the lowest level in a hierarchy of larger political bodies; the highest level is the members of the Politburo (the national level of the Party), whose head is President Raul Castro.

About one-third of all Tutaños are official militants of the Party. These are people who have gone through the intense communist indoctrination programme (as opposed to the general Marxist-Leninist education that all Cubans are expected to receive), and who have the privilege of holding Cuban Communist Party identity cards. All youth and adult members of the PCC must attend meetings held at least once a month (outside work hours) to discuss different 'points' or concerns regarding the way people are behaving within each party member's 'radius of action': an area of about a square mile that surrounds the work unit, usually including at least one neighbourhood. At these meetings, PCC militants discuss whether workers in their work nuclei are meeting the plan for production (as devised by the same Party members) and consider social problems, such as a militant missing a meeting or a worker not showing enough enthusiasm *either* at work *or* in their household (presumably because housework is considered real work, as it ensures the reproduction of Cuban society).

Resolutions 13 and 297 of the PCC, which were passed in 2006 and 2007 respectively, state that even 'simple workers' (those not affiliated with the Party) must participate at least once every 3 months in meetings held by the communist work nucleus that directs their work unit (*Actas de Reuniones* 2007: 14). When I asked a friend, the general secretary (communist leader) of her work unit how simple workers are summoned to meetings, she replied:

GENERAL SEC: The work nucleus tells the administrator [*jefe* of the work
                unit] to invite the cook, for example, to the next reunion.
M: And if they don't come?
GENERAL SEC: They always come.

The 'problems' addressed in these meetings are ordered as a list of numbered agreements, for example a sanction for a worker not performing well in his role. The minutes of these meetings are passed on to the municipal level of the PCC, who is in charge of composing a monthly report called the '*estado de opinion*' (state of opinion): 'based on what is heard on the street, in the market, or from "anonymous" people' (Roman 2003: 92) about possible counter-revolutionary activity. *Estado de opiniones* and minutes of municipal level meetings of the PCC are then sent on to the upper levels of the Party, the final of which is the national Politburo. 'Anonymous' people who offer information for the *estados de opiniones* are usually members of mass organizations, which include *Comité para la Defensa de la Revolución* (Committees for the Defense of the Revolution, CDRs; more on these below), as well as the *Unión de Mujeres Cubanas* (Union of Cuban Women, UMC), the *Asociación Nacional de Agricultores Pequeños* (National Association of Small Farmers, ANAP) and the Committee of Cuban Workers (CTC). Many Cubans are members of such organizations, often because, as one informant told me, it is 'the only thing to do'. As with monthly meetings organized by their work nuclei, workers must attend monthly syndicate meetings, which are mandatory. Workers' mandates collected at monthly syndicate meetings exemplify the way democratic centralism is institutionalized in Cuba. In line with the theory of democratic centralism, the 'integration' of all members of Cuban society into syndicates and other mass organizations is a prerequisite for civic representation.

In reality, the ideal social contract between state and workers, founded on the principle of democratic centralism, establishes a scalar hierarchy encompassed by the national communist party and the president of the republic.

The scalar politics of the Party is, in turn, tied to a temporal imaginary. Thus the Revolution is depicted as an 'historical' process, even from the standpoint of the present (see the official definition of the Revolution, provided by Figure 5.1). In one of the latest amendments to the constitution, socialism is treated as 'permanent and irrevocable' (Constitución de la República de Cuba 2003 [1992]), suspending in time the fundamental political relation between the Party, the state and Cuban citizens.

As I argued in the previous chapter, recent plans to rationalize the state workforce in Cuba do not entirely overturn this transcendental idea of socialism. Indeed, the increase in private enterprise has emerged in the context of a reinforced ideological commitment to a 'sustainable socialism', exemplified not only by policies aimed towards the 'just' re-distribution of profits through taxes and mandatory contributions but also by the government's recent commitment to agroecological production for the domestic food basket (see chapter 6). Justifications for increasing private enterprise and, in some cases, private property (particularly in the agricultural, tourism and mineral sectors) mirror traditional justifications for the advantages of those with political capital. Indeed, as I will explain, the privileges of people with political capital are also justified in terms of how much they give back to local and national communities.

Citizens are shown how to 'follow the line' in both formal meetings and assemblies and via informal interactions with their superiors at work, in their neighbourhoods (through the local CDR) or from other members of the communist party. All actions considered counter-revolutionary – where self-interest overrides collective interest – are dealt with either on a one-to-one basis or through institutions for surveillance which aim to 'protect' Cuba and its citizens from self-interested 'outsiders'. Individual Tutaños are 'oriented' and 'protected' through moral and economic controls, such as those over petty traders of commodities or hard currency. On an everyday level, a person whose primary earnings come from arbitrage or speculation (buying cheap and selling dear) is considered both lazy and counter-revolutionary. For instance, the emergence of illegal money changers during the 2005–2007 period led many Tutaños to comment about the laziness of these 'people who do not work ... carrying wads of money in their pockets'. Any Tutaño of working age who acts solely as an intermediary or illegal petty trader is suspected of anti-revolutionary behaviour, placing profit above communal interests.

In the name of national interests and 'protection', all Tutaños are encouraged to tell the authorities when they see counter-revolutionary activity, though in the case of money changers some illegal activity is permitted by necessity. The following poster on the wall of the Centre for Consumer Registry, the place where ration cards are made and distributed, illustrates the call for all members of the local community to help 'protect' society from outsiders:

Unique System of Exploration (Sistema Única de la Exploración; SUE):

When you hear or see what you suppose to be an Activity of the Enemy or suspect that it may affect the interests of our working people, immediately inform with the following phone numbers: [list of numbers]. We do not underestimate any detail that you are not accustomed to seeing, no matter how insignificant.

YOUR REVOLUTIONARY DUTY: TO BE VIGILANT AND INFORM

The Council of Municipal Defense, Tuta (my translation)

As this poster reveals, the Council of Defense and all representatives of the Party should be notified about the actions of outsiders *within* who have different 'interests' from the goals of the Revolution. As we shall see, Tutaños used the same word – 'interests' – to differentiate insiders from outsiders within their own 'imagined' communities.

Local informers have the job of differentiating real Cuban citizens from traitors. In accordance with the common Cuban expression, '*pueblo chiquitico, infierno grande*' (small town, large fire), informers cause much trouble in small localities such as Tuta. As one Tutaño put it, 'little Fidels' claim that 'there were two bottles [of rum] when there was, in fact, only one'. Some Tutaños told me that because of this high level of surveillance, '*todo el mundo tiene que seguir la misma linea*' (all the world has to follow the same path). Another close friend said:

> Nothing goes unsighted in Cuba. If you commit a 'crime' in your neighbourhood or your workplace [such as not completing the plan instituted by the superior level], the news will spread to the municipal, provincial and then national levels. All actions that do not accord with the Revolution are listed in your *dossier*, of which even Raul has a copy.

It is the responsibility of all militants to observe and report any counter-revolutionary activity within their 'radius of action'. If a militant does encounter negative activity, they are to confront the individual(s) at fault personally and 'fraternally' (Cuban Communist Party 1999: 24; Unión de Jóvenes Comunistas 1989: 2):

> If in one work centre the main problem is the poor quality of products made at that centre or the increase of workers' absences and the loss of work days, the Party's leadership must control this situation, ... applying methods so that they can rapidly find solutions to these deficiencies and persevering until arriving at the objective, combining the demands and help from responsible *jefes* with direct political-ideological labour, man-to-man [interaction with] everyone who is related to the problem in question. (Cuban Communist Party 1999: 16–17; my translation)

Ironically, hierarchical distinctions between people that 'walk the front of the line' and others who must follow are essential in validating the unified nature of the Cuban revolution, for members of the communist party are expected to work in the public interest. The unification of public and private interests is a central tenet of democratic centralism, which is based on the equal allotment of collective property to all citizens. As in other countries, however, the equality of citizens in Cuba is premised on an uncertainty of *who* qualifies to be a citizen in the first place: 'Suspicion ... occupies the space between the law and its application. In that sense, all judicial and policing systems of the modern state presuppose organized suspicion, incorporate margins of uncertainty'(Asad 2004: 285).

Traversing public and private in everyday life

Interconnections between the individual and the collective in Cuba stem not only from the centralized political economy, but also from practical difficulties of living in a world with scarcities and uneven access. Given the need to 'resolve' (*risolver*) at least some problems by soliciting the help of others, public life and private life are often connected by necessity. For instance, the *jefe* of a *bodega* (rationing centre) in Tuta told me that one of his workers was a woman who had recently separated from her abusive husband. The woman expressed to the *jefe* her concern for how she was to find transport to and from Tuta after work in the evenings, now that she was living with her mother in a nearby town, called Barbudo (again, a pseudonym). Barbudo is a municipality located only a short distance from Tuta by car, but it is a significant stretch when one has to rely solely on public transport, which is essentially non-existent in the evenings. The woman told her manager that she was willing to risk being beaten again, going back to her husband so that she could be in Tuta to work. Her *jefe* responded that she should not go back to her husband, that he would find a solution to the problem of transport and, through his connections (which were, revealingly, left out of the discussion), eventually helped her gain access to a small apartment in Tuta. If the transportation system were like it is in Europe or the United States, the worker would never have needed to bring family issues into her work relationship with her boss.

The connection between public and private spaces is also exemplified by close relations between neighbourhood CDRs and households. According to a president of a local CDR: 'Nearly everyone is a member of a CDR in Tuta, so all have the job of being vigilant ... keep[ing] our eyes open so they do not take us when our backs our turned!' According to my experience living in a household, the category of *cederistas* is separate from that of local informers. The latter are more likely to get people into trouble while the former are 'everyone'. The informers in our neighbourhood were pointed out to me from the

Figure 5.3   After the Special Period, hot and crowded lorries (*camiones*) became a common method of public transport. This one is driven by a *particular*, a self-employed driver whose truck is state-owned (like most *camiones*). Source: author.

first week of fieldwork, information that was particularly relevant when my partner (who did not know local rules of conspicuous consumption) came to visit during fieldwork, as illustrated later by an anecdote about beer. Since 'everyone' is a *cederista*, all must be involved in public activities planned by the neighbourhood committee, such as volunteering to help move old refrigerators for new ones as part of the 'Energetic Revolution' programme that was underway during the fieldwork period. Moreover, *cederistas* are expected to offer goods or services to neighbours in need on a periodic basis.

As members of their local CDR, Tutaños are aware of the moral and political obligation to be openly generous and sociable, at least with everyone in their neighbourhood. As I have written elsewhere, there seems to be 'a fine line between true solidarity among neighbours and the political requirement of helping neighbours when in need' (Wilson 2012: 284). In this context, I told the story of a time when tensions between neighbours over exchange relations were felt strongly in the household:

> I heard Jorge and Clara whispering excitedly about something in the kitchen. Because the two are as close as family, I asked them what they were talking about. Clara responded: 'We are talking about María [the next door neighbour]. She asks for everything! Today she asked for sugar, which I had to *buy* [usually enough sugar is available via the neighbourhood *bodega* where state quotas of cheap foodstuffs are distributed]. ... They are asking us for things but they are better off [financially] than us!'

Figure 5.4   Cuban children helping an elderly man down some steps. Source: author.

Despite their hidden annoyance, my Cuban parents ended up giving María a
bit of sugar. (Wilson 2012: 284)

Public pressure to be sociable and generous directs some neighbourly
actions, though others reflect a cultural solidarity that may not stem
from conscious political obligation (Figure 5.4). When Clara was acci-
dentally hit with a baseball bat by one of a few careless boys playing in
the street, the entire neighbourhood came over to see if she was alright.
For the next few days, people constantly came to the house to bring
foodstuffs or check that she was doing better and offer help, using the
common expression '*cualque cosa, me avisa*' (whatever you need, let me
know). We shall see similar associations between political obligation and
cultural practice in the following section, which explains normative hierarchies of 'culture' and 'family' in Cuba. Understanding these will be
useful later on, when I consider how such representations are used in
practice to differentiate acceptable from unacceptable privileges and
exchanges in Tuta.

Levels of culture model and the nation as a family

Like social actors who use 'the law' to legitimize self-interested actions, the state, conversely, uses a language of kin, family, and body to lend immediacy to its pronouncements. (Michael Herzfeld 2005a: 2)

As Raymond Firth (1959) argued for the Tikopia, so for Cubans, categorical systems change with new historical and political economic circumstances. Moreover, cultural categories often shift *in negation* to prior categories. Thus while in pre-1959 Cuba 'culture' referred to a consumer culture that in many respects resembled its North American neighbour (as illustrated in chapter 2), in revolutionary Cuba this word means just the opposite. Rather than tied to individualized property, in post-1959 Cuba both 'culture' and the 'family' have been linked to the supreme value of sacrificing one's own interests for the patrimony of the Cuban nation.

In China, the model of 'virtuocracy' (Shirk 1984) was an alternative to both the hierarchy of privilege in the state arena and inequalities created by market capitalist relations. China's implicit policy 'rejected system integration either through the medium of money or that of power' (Stockman 1992: 262, 273). The virtuous person in the Chinese model of virtuocracy is comparable to the person with a high level of 'culture' in Cuba: one who places the interests of the national 'family' above their own interests and even above those of his or her close friends and family. Similar to the model of democratic centralism explained above, in Sergei Nechaev's idea of the professional revolutionary (and that of most other radical or millenarian movements) equality in a revolutionary society is premised on the idea that the national (or universal) family should come before any personal ties (Nechaev discussed in Wolf 1969: 79–82). Thus according to Nechaev, the true revolutionary is

A man set apart. He has no personal interests, no emotions, no attachments; he has no personal property, not even a name. Everything in him is absorbed by the one exclusive interest, the one thought, one single passion – the Revolution. … All the gentle and enfeebling sentiments of kinship, love, gratitude and even honor must be suppressed in him by the single cold passion for Revolution. … Revolutionary fervor has become an everyday habit with him, but it must always be combined with cold calculation. At all times and everywhere he must do what the interest of the Revolution demands, irrespective of his own personal inclinations. (Prawdin 1961, quoted by Wolf 1969: 81)

Nechaev's idea of the professional revolutionary may be compared to the revolutionary devotion of Fidel Castro and other personified icons in Cuba, such as Ernesto Che Guevara. For some Tutaños, Fidel Castro represents the ideal person whom everyone in Cuban society should strive to emulate.

One Tutaño who identified himself as a revolutionary (though he was not a party militant) compared Castro, whom he saw as 'represent[ing] all the people', to local *jefes* who 'think only about themselves and their friends and family'. According to his account, which is not far from the main 'line' of the Party, while local *jefes* may partake in '*amiguismo*', or the use of power and institutional access to help friends and family, Fidel 'has no friends, he has no family. ... Fidel does not treat any one person better than another'. Indeed, as indicated in chapter 2, to prove that his solidarity rested with the nation rather than his own relations, Fidel Castro destroyed his own family's lands in the early years of the Revolution as 'an example of economic warfare against the rich of Cuba' (Szulc 1989 [1986]: 10).

Like Fidel Castro, Ernesto Che Guevara was a national hero because he placed the politics of the 'universal' revolution above his personal relationships. In a letter to his mother, Celia, written in early July 1959, Guevara negates family ties altogether (though at that time he was a son, a husband and a father):

> I have no house, wife, children, parents, or brothers; my friends are friends as long as they think like me, politically, and yet I am happy, I feel important in my life – not only a powerful inner strength, which I always felt, but also an ability to influence others and an absolutely fatalistic sense of my mission which frees me from all fear. (Castañeda 1998: 166)

Before the needs of the national family could dominate over all others, drastic differentiation between nuclear families had to be eliminated in Cuba. Most upper and middle class families left the country in the first years of the Revolution; their property was then re-distributed to poor families through institutions such as the Ministry of the Recovery of Misappropriated Goods (established in February 1960; Valdés 2003: 91). After 1959, all public utilities and private industries were nationalized, and a wage scale was set whereby the wages of the poorest and wealthiest workers differed only by a ratio of 1 : 5 (Valdés 2003: 21). Money did not matter as much as it had done, just as Castro and Guevara had wanted.

The shift to a centralized, paternalistic system seemed to affect local moralities even then. Elderly informants I interviewed saw a big difference in personal morality before and after the Revolution: 'The Revolution taught people to live and how to use money'. As money became partly a token exchanged for work experience and devotion to the Revolution, paternal heads were called to stop using it for frivolous desires such as drink, and to start buying needed provisions for the family. The exchange value of commodities, such as rum, was overshadowed by the use value of household provisions, and the private realm of the individual was ideally inseparable from the public and economic realm of social property. Though never fully realized, the model 'To each according to his capacity, from each according to his ability' coincided with the idea that provisions for the household (as well as construction materials for the house) were *the* symbol of collective wealth guaranteed by the re-distribution of national social property.

At present, legal preference for national social property over individual property is evident for example in post-mortem hereditary rights over farm-land. Inheritors are evaluated in terms of virtue or revolutionary culture as opposed to how closely they are related to the deceased. Priority is given to those who will work the land to increase state property, as I discovered in a conversation with a former administrative assistant for the municipal level of ANAP, whose job it was to send cases of disputes among *campesinos* to the provincial lawyer and president of ANAP. She told the story of a recent quarrel between two brothers about what to do with their father's land after he died suddenly. The elder brother wanted to sell the land, the younger wanted to keep it as a farm. The younger won the dispute because it was thought that he had more revolutionary culture. As the former ANAP employee explained, culture in this sense means that he 'sacrificed for the Revolution' by working the land for 'social consumption [i.e. day care centers, hospitals, schools, etc.]', while his brother only wanted to profit from the land. She summed up the law when she said that the elder brother was heir 'simply on paper': he was not heir by his revolutionary conviction and dedication.

Just as the concept of democratic centralism implies a hierarchy, so the idea of culture binds Cuba together into a group with shared assets, like a family, in which some people merit more than others but all have a share in some of the goods (as opposed to non-family members). In the official value system, culture works as a measuring rod for estimating revolutionary dedication and selflessness. At the scale of the everyday, culture is also associated with a lack of interest in self-gain, as illustrated by the common expression: '*el que mucho abarca, poco aprieta*' (one who tries to get a lot of things, ends up with very little).

The value of revolutionary culture also has a racial dimension. For instance, during fieldwork Afro-Cubans were regarded by many as being more 'interested' (in my money and/or sexually) than others with higher levels of culture. The levels of culture model departs from the pre-1959 idea of culture in that it associates self-interest and individualized consumption with negative vices, but it shares with its earlier counterpart the bias for whiter skin, as well as a polished appearance and an educated outlook. Similar to the Bolshevik notion '*kultura*' (Scott 1998: 195), the 'new man' in Cuban society must be both morally sound on the inside and *visibly* acceptable on the outside. Indeed, through representations like revolutionary culture, the very body becomes a scale at which relations between internal and external are subjected to judgement. As Mette Louise Berg writes:

> The concept of 'cultural level' [in Cuba] is related to a host of indicators including an ethos of moderation, decency and restraint such as not raising

one's voice in public and abstaining from public drunkenness. ... Roaming the streets in search of tourists clearly does not conform to having a 'high cultural level'. (Berg 2004: 52)

Nadine Fernández, another anthropologist of Cuba, places more emphasis on the Cuban idea of culture as hierarchical: 'Culture [in Cuba]... implies a social hierarchy: one group with high culture (*más nivel, alta cultura*) and the other group with low culture (*menos nivel, bajo* [sic] *cultura*) ... [it is the] level of formal education, public manners and etiquette, and ... degree of social refinement' (quoted by Roland 2006: 155). But Fernández's definition of culture in Cuba could fit any society which employs similar status hierarchies, such as those present in pre-1959 Cuba or even in the United States, where people with money to buy tablet computers are often considered superior to those without the latest gadgets who use ones like the old computer on which I write. What makes the concept of culture different in present-day Cuba is its setting within a communist system, according to which those seen to be upholding revolutionary values – as opposed to commodity values – are justified as more deserving than others.

Shifting scales of appropriate exchange

Spaces and exchanges in between state and civil society: the case of jefes

We have seen that militants of the Party are officially considered those who must lead the people. When I asked a woman what being a militant meant to her, she replied that it meant to 'walk at the front of the line'. *Jefes* (who are usually members of the Party) and others recognized as leaders may 'walk at the front of the line' in material terms as well. Indeed, the authority of *jefes* to ensure that workers reach production targets (what Norman Stockman calls 'one-man management' in the context of Chinese communism; Stockman 1992: 262–3), often means that managers have sole control over the final product (which they may divert to some workers), though the latter is supposed to be passed on to the national scale and then re-distributed as social property. The PCC attempts to control all acts against the Revolution in the workplace, as well as the general issue of *amiguismo* (also called *sociolismo*): diverting national property or positions of merit to family members and friends (comparable to *blat* in the former Soviet Union). Indeed, the *Rules of the Communist Party* (Cuban Communist Party 1999: 35, 40, 63–6) repeatedly states that all members of the Party must 'fight' against *amiguismo* and other violations of revolutionary values. But despite the efforts of the Party to limit the reaches of political capital, the number of *jefes* who use this method to redirect national property to themselves and to their *familiares* has increased since the Special Period, representing a shift in scale from the 1980s when provisions were not as difficult to obtain from state sources as they are at present.

Like that of *particulares*, to whom I will soon return, the privileged status of state managers leads to moral dilemmas at the local level concerning the social balance between revolutionary culture and personal benefits. In local terms, people 'with culture' do not take advantage of their position to the detriment of others, though they may engage in illegal activities such as taking state property from their workplace. Thus Tutaños differentiated 'good' *jefes* from 'bad' *jefes* according to whether items taken from the state were destined for collective benefit or for personal gain. 'Grabbing' (demonstrated with a grabbing gesture rather than verbalized) things from the state for one's family or for a wider field of local 'kin' was often contrasted with 'stealing' goods from the state for profit. Although the law does not differentiate stealing from 'grabbing' social property – both are illegal – the latter gesture indicates an illegally licit activity, a way of justifying an unlawful act in social terms.[3] The grabbing gesture is a locally acceptable idiom for using state perks to support one's *familiares* as opposed to benefitting individually. Of course, a concept like *familiar* has no boundaries, and I have heard Tutaños complain that *jefes* help 'not only their families, but the families of their families, and *their* families and *their* families, and so on.'[4] Despite the differentiation between family, extended family and *very* extended 'family', the normative distinction between provisioning for one's *familiares* and profiting individually is important in Tuta.

Because of the shared 'cultural intimacy' (Herzfeld 2005a) between ordinary people and those in power – in the case of Tuta, the universal need to provision food for the family – state functionaries as well as ordinary people perform illegal acts. In one of our informal interviews, a vendor of the free farmers' market (in which produce is theoretically sold according to the laws of supply and demand) told me that prices in the market are not the same as prices the farmers charge. As the man, Angelo, said, 'the vendors in the *mercado agropecuario campesino* do not produce every item they sell in the market'. At least during the research period, selling goods not produced in the household was a violation of a law requiring all vendors in the free farmers' market to sell only what they or their family members produce on their farms. This law was implemented to ensure that an increment in price would not result from the use of intermediaries. Despite the law (eased in 2011, at least for the non-farming sector), every day at 7 or 8 in the morning a lorry filled with produce parked near the market and sold fruits, vegetables and tubers to vendors at inflated prices. Angelo explained how this violation of the law affects prices:

Figure 5.5   A vendor at the farmers' market, one who is also a farmer. Source: author.

This man buys 80 pounds of tomatoes at 2 pesos a pound from a farmer, sells the whole lot to farmer/vendors at the *mercado* at 3 pesos a pound, and the farmer/vendors sell the tomatoes to the population at 4 pesos a pound. That is only 2 tomatoes for 4 pesos! And the majority of the population who make 10 pesos per day suffer. Because of this situation, everyone in Cuba is worried about the question, 'What are we going to eat today'?

I asked Angelo how vendors at the free farmers' market get away with buying goods they do not produce from intermediaries and selling them illegally at such inflated prices, when I assumed there were state inspectors who would fine them for this. He laughed and said:

The inspectors are in on it. ... as well as the *jefe* of the *mercado agropecuario campesino* in Tuta. Each vendor gives the latter five pesos per day to close his mouth [amounting to about sixty pesos per day or $1/£1.50]. And when inspectors come, vendors tell them, 'meet me over at the corner in ten minutes'. And they [the inspectors] do and receive bribes ... Meat vendors are also ready for the inspectors. They make a special bag [of meat] for them so that they are ready whenever the inspectors make their appearance.

Angelo concluded with what he thought to be a necessary disclaimer: 'But all of this is not the state's fault. Fidel, Raul and the members of the Politburo have good intentions but they do not know what is going on. It is the fault of individuals, not them'. As this last comment illustrates, in the 'local' moral economy, negative occurrences are often seen as the fault of individuals as opposed to actions at the national level ('They'). At least for some Tutaños, the presence of perhaps more than a few free riders at the bottom does not undermine the legitimacy of those at the top.

The ideal relation between state representatives and civil society is regularly muddled in practice, for most militants and *jefes* are not cold bureaucrats whom one never meets but one's neighbour or even a close relative. Like other countries, in Cuba there is no strict divide between 'highly personalized forms of private power and the supposedly impersonal or neutral authority of the state' (Das and Poole 2004: 13). Similar to more recent efforts in geography to 'assign a home to globalization' (Marston *et al.* 2005: 427), anthropologists have argued that there is 'no necessary site for the state' (Trouillot 2001: 126). The 'Russian doll' approach which places the state (or the globe) at the top of a hierarchy of scales should not obscure 'the site of ordering practices' (Marston *et al.* 2005: 427), which do not work at any predetermined scale. Just as one cannot make a strict separation between global economic 'spaces' and everyday 'places', a causal relation between regulation from above ('the state') and opposition from below ('civil society') is too simplistic. Such a perspective counters the general tendency in the social sciences to romanticize resistance as a strict negation of bureaucratic or economic power, sometimes attributing all meaningful politics to the local scale (Amin *et al.* 2003). As Marston *et al.* (2005: 427) argue, blaming all social ills on 'macro' forces like globalization or the state 'obscures those sites of ordering practices, as well as the possibilities for undoing them'.

In reality, ordinary people in many places subsist by collaborating with state functionaries for their mutual benefit, as Herzfeld (2005b) argues for the Greek case:

> The fact that tax would often be calculated on the assumption that the parties were declaring only half of the true value of the goods transferred – an assumption in which the parties naturally enough concurred – demonstrates the complicity of bureaucrats with their clients as well as the practical compatibility of a precise idiom of documentation with some deliberately very imprecise forms of financial agreement. (Herzfeld 2005b: 371)

Herzfeld's account is far from an isolated case. In communist China, for instance, local cadres who are supposed to implement the one-child policy have to balance official rules with local norms (e.g. one *male* child policy). 'At the bottom rungs of the state apparatus, township and village cadres were predisposed by cultural values and social obligations to soften state policies to accommodate pressing peasant demands' (Greenhalgh 1994: 13). Empirically, therefore, the normative power of some scales over others is negotiated at the scale of interpersonal, everyday relations.

Cuban *jefes* who benefit from bribery or take provisions from their workplace demonstrate perhaps most clearly that the 'Fight' of all Cubans to protect national property is complicated by an everyday 'fight' for household sufficiency. As one Tutaño put it, mirroring another's comments cited in the introduction, 'Everyone wants a top state job now only because they want access to state goods. Before [the Special Period], it was not like this. No one stole from the state, they were morally against this. Now everyone steals from the state'. Privileges that emerge through illegal activities do not necessarily imply extreme differences between the 'haves' and the 'have-nots'. 'Everyone in the world' (*todo el mundo*) prefers to buy goods sold at lower prices from illegal street vendors who do not pay taxes rather than from the '*Bandido del Rio Frio*' (Cold River Bandit), a phrase used in reference to the high-priced farmers' markets, as noted in chapter 3.

An example of a good *jefe* is Jorge (my Cuban 'father'), a quiet man in his mid-sixties whose sinewy yet strong arms give the impression of someone who has worked very hard all his life. One day, Jorge explained to me that he had no problem 'grabbing' (he communicated this concept with a gesture — one hand clutching the air) avocados, chillies, plantains or other produce from the state cooperative where he had once administered as *jefe* but which had since been abandoned. When Jorge started as a worker in the cooperative, he 'took things like everybody else'. But when he was elected president of the cooperative by the agricultural workers' syndicate, he stopped stealing and started to sell more produce at the farmers' market. This was surplus left over after covering the plan for *autoconsumo* (production to feed the cooperative workers' families or for workers' lunches). 'Everyone in the world bought from me. My goods were the cheapest as I brought them directly from the fields, eliminating the middle-man'.

When the *jefe* of all the *autoconsumo* areas of the cooperative took over Jorge's job, 'he made a million pesos one year and it was all lost'. I asked him how this happened. 'It was all stolen. The *jefe* took it all. He was there for only four years and the cooperative had to close because of a broken turbine, which could have been fixed with the money he stole'. He criticized the 'bad' *jefe* for 'stealing' provisions that Tutaños could have continued to buy at the market, had Jorge still been in charge.

J: *No tienen vergüenza* [they have no shame]. When they put in another *jefe*, at
       first the workers liked him. But after a while, they realized that he was
       pinching things for his own pocket, and [the *jefe*] wanted the workers to
       work harder! They knew what he was about so they sat and did not work.

M: And why did you get fired, when people knew that the *jefe* who replaced you was no good?

J: Because when I wanted to go to work with the tractor he would say: 'The tractor is broken', and it wasn't. He would tell the others that there wasn't any gasoline. And when they fired me, I told the superior *jefes* that the fall in production was not because of me – they thought it was my fault – but because he was stealing petroleum from the state which was supposed to be for the tractors. He also went to the cooperative on Sundays when no one was working and took five pounds of *yuca* [cassava] from the cooperative. He was supposed to bring the *yuca* to the primary school nearby, but only gave them two pounds and took the rest, selling the remaining three pounds in Calle 'X'. I told all of this to the *jefes* of the cooperative and to higher *jefes*, and no one did anything. I didn't get my job back. Do you know why? *Because they were all stealing from the cooperatives.*

According to a small group of Tutaños who worked under Jorge at the *Cooperativa de Producción Agrícola* (Agricultural Production Cooperative, CPA), whom I subsequently interviewed, Jorge was a 'good' *jefe* because he sacrificed his own interests for that of his workers and the whole community. They distinguished Jorge's use of power to help others obtain provisions from the 'interested' acts of the 'bad' *jefe* who took over his job at the cooperative and caused it to become unproductive. While Jorge only 'grabbed' a few goods from the cooperative for his *familiares*, and, moreover, provided inexpensive provisions for the whole community, the new *jefe* 'stole' hoards of produce to sell on the street for his own profit.

People like Jorge only take from the state when they feel the latter is not providing enough for their family or community. Alexei Yurchak used the phrase 'entrepreneurial governmentality' in reference to a similar way of using state privileges in the Soviet Union (cited by Wanner 2005: 526). Such a reformulation of social justice is evidenced by the story of another man, Angelo (mentioned above), who told me how he provisioned a few inexpensive chickens by going 'to the left (*de la izquierda*)', that is, into the illegal market. Usually live or dead whole chickens, though thin, cost over 100 pesos (4 Cuban Convertible Pesos or CUCs), but these cost only 40 pesos each. When I asked where Angelo obtained the chickens, he spoke openly about the acquaintance from whom he buys chickens illegally. According to Angelo, who considers himself a revolutionary (though he was not attached to the Party), his acquaintance discreetly 'grabbed' (again, the gesture) a few chickens from his *granja avícola* (state poultry farm) each day after work, quickly killing the chickens by wringing their necks and placing them in a large, opaque sack so that no one on the lorry home knew what he was carrying (the use of opaque bags is a common tactic for transporting illegal goods in Cuba). Angelo said of the man that 'he is just defending his *rights*. He does not get enough [salary] from the farm so he must take chickens and sell them'.

Jorge's defence of 'grabbing' produce from the *campesino* for whom he worked in 2007 was similar:

> This campesino is *ruin* [stingy]! He has no culture! I only get what I grab from him [gesture] ... He doesn't pay me for the weeks I don't work Saturday – and I mostly come on Saturdays. ... He gave me a pumpkin that he would not even have given his animals! I was too embarrassed not to accept it. I will get a better pumpkin this week [grabbing gesture], so that we can have it for our *caldo de viandas* [tuber stew].

Normative distinctions between emerging entrepreneurs in Tuta

Most Tutaños who choose to enter into the illegal economy do so not to fan anti-revolutionary sentiments, but rather to make ends meet in a world where alternatives to the state distributive economy are increasingly necessary. Indeed, those who 'grab' what they feel is their due often defend their actions by using similar values set forth by the Revolution: social justice, self-sacrifice, community, and so on. Ordinary people (i.e. not *jefes*) who must engage in the illegal exchange of foodstuffs also justify their actions by reworking existing revolutionary values in new economic and spatial contexts. For instance, Clara (my Cuban 'mother') sold spaghetti to neighbours that she bought in bulk from a middleman, who in turn bought it from another. The latter was a state worker who processed the spaghetti from that left over from a state factory where better-quality spaghetti was sold in CUCs (note that this worker was not 'stealing' the best social property, only 'grabbing' the rejected quantity). Clara stressed that her neighbourhood market for spaghetti did not bring her any profit; it only provided enough to buy CUC spaghetti to make up for the lack of state-subsidized spaghetti. As with other products sold illegally on the street that come 'from the factory' Clara's spaghetti was of better quality than what may have been available via the state re-distributive structure. Thus Clara claimed she was also providing a service for people in the neighbourhood who did not have the money to buy spaghetti at the hard currency store. As her story demonstrates, entrepreneurial activity that shifts the scale of provisioning from the state to the neighbourhood is often justified in terms of social rather than individual interests.

A similar example involves the difficult situation I faced one day when a seemingly nice couple invited me to their home for a drink. After meeting Pablo and his *mujer* (woman), we started walking towards their house, stopping to let Pablo buy rum on the way (as a test of character, I allowed him to make the purchase without offering to buy anything – I had been constantly warned about Cubans who only spend time with foreigners to eat and drink for free). While Pablo bought rum at a small booth next to a state-owned restaurant, my new friend, Julia, who worked at the CUC store across the street, called me over. I went to talk to her and she whispered in my ear: 'Those people are dangerous. ... They have no culture, do not spend time with them. Do not *be seen* going to their house'. After her warning I noted that almost all the people on the street were watching me walk with Pablo and his *mujer*. Though I had become accustomed (as well as one can be) to stares and *piropos* (sexual comments from groups of men who gather at street corners or sit in open areas), these looks seemed different in nature; they expressed what seemed to be a mix of apprehension and near-mockery.

Despite the advice from my friend (or perhaps because of it?), I did eventually go to the couple's home. Keen in my role as a socially awkward ethnographer, I was curious to find out what my 'family' thought about my new acquaintances. After our meeting, I invited Pablo and his *mujer* to walk me to the house. The couple entered in an ashamed fashion without sitting. Pablo tried to explain to Clara that he regarded me with the utmost respect, 'as a father would his child'. I thought this an odd comparison as Pablo was not so much older than myself. The short meeting ended there and Pablo and his *mujer* seemed to be guided out of the house by an invisible force of negativity on Clara's part. After the encounter, Clara told me that these people were no good, that they 'have no culture', that they 'steal things from *everyone*'. The next day, I asked my friend, Julia, who had initially warned me about the couple, to explain why Pablo and his woman were such outcasts. She gave a similar story to Clara's: 'These people steal from *individuals*, not just from the state. They are known as people who place their own self-interests above the needs of their fellow "comrades". They steal bikes from our people and sell them for profit'.

I was to find out later that it was 'dangerous' to spend time with people who had 'no culture', as being seen spending time with them posed a threat to both my 'family' *and* to my research plans. As a friend from Havana told me: 'Other foreigners have been deported for doing such things!' This 'danger', which seemed especially threatening to foreigners like myself, was measured in terms of whether a person was considered with or without culture or 'interested', as further illustrated later in the chapter.

Insiders and outsiders in the market sphere: particulares *and* familiares

Particulares do not create social property for the nation which is then re-distributed to households as use value; instead, they create property within their homes or 'grab' or 'steal' it from their state workplaces, gaining from exchange value. In the post-1990 period at least, the *particular* has, in part, taken over the state's role as provider. It is precisely for this reason that it is one of the ambiguous categories of persons that must be re-defined in both national and local terms.

As opposed to formal exchanges between, for example, state and workers, exchanges of information and foodstuffs do not take on any particular structure (Foster 1961: 1186) and it is common for complete strangers to ask each other for help in finding a scarce item. This sociability of provisioning also breaks down the divide between state and civil society, as illustrated earlier. When one day in the neighbourhood the usual street vendors and displays of produce for sale on front porches vanished, I was told that 'everyone' knew that inspectors were in the street who would fine illegal vendors. 'How did people know this?' I asked. 'Members of the local tax office told their friends and families, who told others, who [then] told us'.

The link between *what* one buys and *who* one buys from was evident at the next door neighbour's house when I went there to trade 20 CUCs for 480 pesos, a favour which Marta (the neighbour) had asked of me earlier that day. The woman who had just sold us fish at Clara and Jorge's house was talking to Marta and playing with her 1-year-old daughter. The fish was cold, fried and not looking delicious, but Clara bought it anyway as the woman was a trusted *particular*. While Jorge and Clara are most often very sceptical of buying fish – lately there had been sellers of low-quality fish posing a risk of sickness – they told me that this 'must be good' fish as they knew the woman vendor personally. That evening we ate it and I was sick with diarrhoea and stomach flu the next day. Clara and Jorge were reluctant to blame it on the fish, especially since no one else became ill, but I suggested that there may have been a few bad pieces in the woman's bucket. I wondered whether they were sceptical of my theory because they were friends with the woman who sold the fish.

When social relations of trust replace a 'generalized trust' in the national currency, inter-personal relations eclipse the 'all-pervasive links of a monetary system' (Humphrey 1998, cited by Trouillot 2001: 128). The criteria of trust in evaluating *particulares* is partly based on values embedded in the national moral economy. For instance, *particulares* who charge what is considered too much money for their commodities or services are referred to as *abusadores* (abusers; Wilson 2012: 284). By contrast, trusted *particulares* engage in the market sphere but do not place their own interests in money over the need to offer foodstuffs at fair prices to the local community.

Contrary to this moral range of *particulares*, the term '*familiar*' usually defines a more intimate social space where 'generalized reciprocity' (Sahlins 2004 [1974]: 195–220) is at play – the long-term exchange of goods. 'If I am missing one or two eggs, I can ask Linda if she has some. If she does, she gives them to me. Another time she is missing flour, and I give her some'. Treating others like family involves the willingness to give goods or services when necessary, even if there is no political requirement to do so. Exchanges of foodstuffs between 'kin' are often of an informal, dyadic (two-way) nature. As opposed to gifts, cooked food is often accepted without thanks, as in other small communities (Foster 1961: 1180). Even if one asks a *familiar* to '*regalarme*' (give me a gift of) a few lemons, this is still a common way of depicting exchanges between two or more people involved in a long-term informal exchange relationship. The value of long-term exchanges of goods or services between *familiares* is most evident in times when the rule is broken: when a friend offered me use of her work computer to send emails and did not offer up this service to another close friend who was in need of it, another *familiar* (Clara) was upset. 'If she has access to email she should have told him about it!' she gossiped angrily.

Contrary to receiving gifts from people 'without culture' who always have other 'intentions' in mind, small items or services are accepted (often reluctantly) either when it is considered a way to thank the person for their help or when it is an open act of generosity between friends and family where no payback is expected. When I knew I would be accepting phone calls at a neighbour's house (the neighbourhood phone) over the fieldwork period, Clara advised me to give a few candy bars of better quality than those available in Cuba (brought from the United States), to the woman of the house: 'Do this just the first time you use the phone, the rest of the time it is unnecessary ... Cubans are like this'.

As in any economy, reciprocal or long-term exchanges of food and other goods in Tuta coexist with short-term exchanges of commodities (Parry and Bloch 1989: 2; for a distinction between 'goods' and 'commodities' see Gregory 1982, 1997). Like other rural societies, household continuity in Tuta depends on a combination of different economic strategies (Long and Richardson 1978). Indeed, long-term exchanges between *familiares* are often predicated on shorter-term exchanges with trusted *particulares* that circulate on a wider scale. For instance, while items exchanged between families are usually cooked food or beverages, the raw materials are often procured through petty commodity activity and may come from outside markets. When members of our household could not locate a vendor from whom to buy local cheese (i.e. cheese made in Cuba), Clara asked family members who asked other *familiares* who asked still others. After several days, a trusted *particular* was finally located who agreed to stop in front of the house one morning per week to sell us cheese that derived from a CUC store in Havana (likely imported) at the same price that Clara's *familiares* paid: 10 pesos per 0.25 lb (0.11 kg). Clara contrasted the trusted vendor with *abusadores* 'who sell the same amount of cheese that quickly expires for 15 or even 20 pesos per liter!'

The category *familiar* not only refers to close people between whom long-term exchanges are transacted, but also to a person who is allowed to enter into the private arena of the house as opposed to the visible arena of the street. *Familiares* can walk into one's house without knocking, and are those with whom one can have in-depth discussions 'behind closed shutters'. As opposed to detached neighbours or fellow members of one's CDR or work unit (who may also become *familiares*, if a close relationship is established), *familiares* are those who can be trusted to keep any information which may affect one's dossier away from 'the ears' of the street (such as buying illegal beef). Through the concept of *familiares*, Tutaños carve out a domestic space that is separate from the state and market spheres.

As evident from these examples and others in the section that follows, people in Cuba borrow representations from the national value system, such as metaphors of kinship, to defend illegal or otherwise unseen activity as morally acceptable. When the boundaries of the national family shrink to fit local-level definitions of kinship, social property of the nation may become social property of a 'family' of individuals. Within illegally licit domains of everyday moral economies, however, social property of the family still overrides individual property or self-interest.

Jineteros: *the dangerous realm of the 'interested'*

As suggested in the introduction to this book, all foreigners who visit Cuba are considered tourists: people with hard currency. While tourism is regarded as a necessary evil, it is also officially sponsored as a collective means to earn hard currency, which brings 'benefits [to] the whole population' (Cuesta Álvarez 2007: 4). The nation accepts tourism as a means of acquiring social property, the kind of property that furthers the long-term reproduction of the social and political economic order. Until very recently, however, individual Cubans have been strictly prohibited from associating with tourists (unless, of course, they are privileged to work in tourism, jobs which usually require political capital). Until April 2008, Cuban citizens were not allowed in most tourist establishments, especially hotels (now they are allowed, but like access to mobile phones and other recently legalized consumer goods, very few can afford to stay or purchase things in hotels). This policy was not to my knowledge explained in writing, but it did happen in practice, especially when Afro-Cubans were seen with white foreigners in places like Havana.

Anthropologist L. Kaifa Roland associates the separation between Cubans (especially black Cubans) and tourists with racial segregation policies in the United States in the mid-20th century (an association I have also made myself). When Roland asked a Cuban friend why such segregation takes place in Havana and elsewhere, the respondent was adamant about the need to 'protect the image of the Revolution abroad'.

> 'Can't you see,' he shouted, 'that the only way to fight the [exportation of negative images] is to prohibit [all Cubans]. No Cubans! No Cubans! *No Cubans!*' ... What felt inconsistent to me was that such language was being used in the name of a revolution that purportedly sought to create an egalitarian Cuba. (Roland 2006: 157)

The first few times I visited Cuba I assumed that the idea of 'protecting' tourists from negative images of the Revolution or Cuba was an official way to mask a very opposite agenda, the 'protection' of Cubans from the influence of capitalist westerners. With hindsight, however, I recognize that it is not tourists that are the primary threat. Instead the moral danger arises from *Cubans* who are not trusted to integrate in the market-based arena of tourism. Indeed, the probable reason for the tourist–Cuban segregation is similar to that between monetary value and revolutionary values: the first is associated with short-term gain for individuals while the second are fundamental to the very reproduction of Cuban society. The official assumption is that relationships between Cubans and tourists are characterized by a mercenary interest on the part of Cubans, types of activity that work in contradistinction to the 'social goals' of the Revolution. Any Cuban affiliating with tourists is immediately associated with one who places his or her own 'interests' over those of the Revolution. According to Roland, who has spent more time analysing tourism in Cuba than has this author,

> because the tourists have spending power, they are (perceived to be) constantly in danger that a member of the comparatively impoverished host population will seek access to that money, whether forcibly or more creatively – and lest we forget, the state seeks as much of the economic capital for itself. With the desire to continually attract more tourists, foreigners are seldom chastised for violating the borders; rather, the local population more frequently reaps the consequences. (Roland 2006: 152)

Consistent with the revolutionary value system, both top-down and bottom-up agents demonize Cubans who hustle or even affiliate with tourists, assuming that *all* relations between Cubans and foreigners are beset by material self-interest. This rule of segregation made it very uncomfortable for me and my Cuban friends during fieldwork, as the reader may well imagine. Living with a family and authorized to do so by a family visa, however, I was often thought to be Cuban-American, particularly if I didn't give it away by opening my mouth (or wearing anything but heels outside the house).

Admittedly, many *jineteros/as* (street hustlers or prostitutes, respectively) do work together to earn profits, especially in tourist areas such as Havana. I found this out when I went to a restaurant with a few (black) Cuban friends in the nation's capital. Though my friends had no 'intentions' of using me for my money, upon our arrival they were immediately asked by the restaurant's *jefe* if he should pay them to start a 'business (*negocio*)' of bringing foreigners into the place to eat. Reiterating a common expression I have heard in other uncomfortable circumstances, my friends told me they 'had to laugh so as not to cry'. One tried to find amusement in the

Figure 5.6   The waterfront (*malecón*) in Havana. Source: author.

juxtaposition between people 'without culture' and our well-educated group of friends, exclaiming: 'So, we have a *jinetera* lawyer in our midst!'

Another example occurred in the 2004–2005 period when I was required to stay in a costly and comparatively palatial academic visitors' home in Havana whilst conducting research with agroecologists from the Agrarian University of Havana. When I asked why I could not stay with Cuban friends instead, the director of the department told me that I must be 'protected' from Cubans who may take advantage of me (sexually) or my money by offerings of dinners or excursions, expecting 'favours' in return. It was indeed commonplace to hear habaneros offering foreigners (including myself) gifts of lobster dinners (procured via illegal outlets) or other delights, in exchange for what is considered a lot of money in Cuba. I never accepted any of these offers, but did encounter similar proposals, especially of services such as washing clothes, from the Cuban women who worked in the visitors' home. One woman warned me 'not to let Rosa wash your clothes'. When I did as an experiment, Rosa asked me nearly every day thereafter to buy her a beer or a can of soda from the CUC store across the street. In such a context it was even harder than usual to distinguish between gifts that were purely generous from those that were entirely self-interested.

In Tuta, where I was likely the only foreigner amongst a large population of Cubans (except for visiting relatives from Miami), warnings from friends to avoid accepting gifts from 'interested' people were even more frequent (but perhaps less warranted). Anthropologist Anna Cristina Pertierra has referred to this Cuban taboo of accepting gifts from foreigners. As opposed to what she calls 'circles of respectability', such as exchanges between households, acquiring gifts from a tourist is not considered a valid way of obtaining goods (Pertierra 2007: 144). A 'good' person in Tuta does not accept offers without thinking about the intentions behind the offer as well as who else may be affected if it is taken up. As members of educational and political institutions advised me that I must stay 'in line' with the Revolution, so non-official persons, such as my Cuban 'father', warned me to stay away from people 'without culture'. In accordance with José Martí's oft-quoted adage 'good intentions are shown through actions and not words', Cubans are always cautious of offers of goods or services, always 'looking beyond the offers to the intentions behind them'. As many Tutaños told me, '*Todo el mundo no es igual*' (Not all people in the world are trustworthy). For this reason, Clara (a white woman) was suspicious when my two friends (mulatto and black, respectively, both in their mid-thirties), once 'boasted' to her about how much they had helped me in the past. Clara warned me about spending time with them (though they are now good friends): '*Muchas veces dicen una cosa y piensan en otra*' (Many times they say one thing and think another). A similar well-known expression that demonstrates this attitude of caution is: '*Del dicho al hecho hay un gran trecho*' (There is a great gap between what is said and what is done).

The need to be recognized as one without 'intentions' was perhaps most evident when my partner visiting from England bought several beers and distributed them amongst his new friends at a CUC store in Tuta. In front of the store are seats where Tutaños sit and eat ice cream, drink soda, rum or beer or, most often, simply chat without consuming anything. When my partner came to the seating area with the beers, the reaction of our friends was far from what he expected. Instead of amiable cheer in response to what would be considered a friendly gesture in England, the Tutaños seemed cautious to accept, and one person even got up and walked away. The latter was a money-changer, seen in that very spot every day engaging in the 'lazy' activity of exchanging pesos for CUCs (or vice versa). With hindsight, I recognize that it was 'dangerous' – especially for this person who was already suspect in society – to be seen accepting gifts from tourists, like my partner. Moreover, I now realize that we were sitting right in front of the PCC headquarters, where a guard stands day and night to watch for counter-revolutionary activity! As with gift and other types of exchange, so with production, official and unofficial spaces coexist and complicate distinctions between market, state and life spheres.

Conclusion

This chapter has provided detailed analysis of political (and economic) hierarchies embedded in Cuban institutions, and offered empirical evidence of how such hierarchies influence social differentiations and uneven access to food and privilege on the ground. By (re)connecting 'is' with 'ought', I have illustrated the relation between ordinary economic geographies in Tuta and overarching cultural politics of Cuban socialism. As the ethnographic data in this and earlier chapters illustrate, economic geographies are comprised of both material realities – the availability and capability of using things like money to get by in life – *and* cultural norms. Shifting relations between economics and culture, which occur both within and between individuals, illustrate what Roger Lee (2006) calls ordinary economic geographies. These involve multiple trajectories that variously combine cultural values and structural economic forces. By contrast, scalar projects like that for the Cuban economy 'mis[s] the inherent complexity of ordinary economies and thereby plac[e] limits on the economic geographical imagination – and hence on the possibilities of political transformation'(Lee 2006: 414).

In Cuba, top-down and bottom-up understandings of acceptable economic exchange are 'but two of a host of refractions of a broadly shared cultural engagement' (Herzfeld 2005a: 3). In assessing the possibilities of ordinary economic geographies, therefore, it is essential to take the 'double-edged nature of scale' (Smith 1992: 78) into account: as both enabling and disabling different forms of value. In other words, we cannot underestimate the normative power of community, which may be enabling as well as disabling.

Notes

1   Roitman (2005: 21).
2   The *Día de los Campesinos* is celebrated annually on 17 May. It symbolizes the *lucha* of several prominent farmers in Cuban history, some of whom perished in their fight against 'enemies' who attempted to stop the 'statization' and collectivization of Cuban farmlands in the early 1960s. The symbolic importance of the agrarian laws passed during this period is explained in chapter 6.
3   See appendix 5 for a list of legal and illegal household food purchases over a 1-week period in 2011.
4   Similar groupings were a common characteristic of eastern European communism. For an example of larger-than-family networks in Poland, see Wedel (1986).

References

Actas de Reuniones. (2007) *Partido Comunista de Cuba,* Tuta. Jan.
Amin, Ash, Angus Cameron and Ray Hudson. (2003) The alterity of the social economy. In *Alternative Economic Spaces,* edited by Andrew Leyshon, Roger Lee and Colin C. Williams. London: Sage Publications, pp. 26–54.
Asad, Talal. (2004) Where are the margins of the state? In *Anthropology in the Margins of the State,* edited by Veena Das and Deborah Poole. Santa Fe: School of American Research Press.
Berg, Mette Louise. (2004) Tourism and the revolutionary new man: the specter of *jineterismo* in late 'Special Period' Cuba, *Focaal: European Journal of Anthropology* 43: 46–55.

Bloch, Maurice. (1989) The symbolism of money in Imerina. In *Money and the Morality of Exchange*, edited by Jonathan Parry and Maurice Bloch. Cambridge: Cambridge University Press, pp. 165–90.

Casteñeda, Jorge G. (1998) *Compañero: the Life and Death of Che Guevara*. New York: Vintage Books.

Constitutión de la República de Cuba. (2003 [1992]) *Gaceta Oficial de la República de Cuba*, Edición Extraordinaria, No. 3, 31 Jan.

Cuban Communist Party. (1999) *Rules of the Cuban Communist Party*. Havana: Editorial Política.

Cuesta Álvarez, Leonardo. (2007) Proyecto para aumentar frutales, *El Habanero*, 19 Jan., 3–4.

Das, Veena and Deborah Poole. (2004) Introduction. In *Anthropology in the Margins of the State*, edited by Veena Das and Deborah Poole. Santa Fe: School of American Research Press, pp. 2–32.

Dumont, Louis. (1980 [1966]) *Homo Hierarchicus: the Caste System and its Implications*. Chicago: University of Chicago Press.

Firth, Raymond. (1959) *Social Change in Tikopia: Re-study of a Polynesian Community after a Generation*. London: George Allen and Unwin.

Foster, George. (1961) The dyadic contract: a model for the social structure of a Mexican village, *American Anthropologist* (N.S.) 63(6): 1173–92.

Foster, George. (1965) Peasant society and the image of the limited good, *American Anthropologist* 67: 293–315.

Gluckman, Max. (1963) Gossip and scandal, *Current Anthropology* 4: 307–16.

Greenhalgh, Susan. (1994) Controlling births and bodies in village China, *American Ethnologist* 21(1): 3–30.

Gregory, Chris. (1982) *Gifts and Commodities*. London: Academic Press.

Gregory, Chris. (1997) *Savage Money: the Anthropology and Politics of Commodity Exchange*. London: Harwood.

Harris, Marvin. (1986) *Good to Eat: Riddles of Food and Culture*. London: Allen and Unwin.

Herzfeld, Michael. (2005a) *Cultural Intimacy: Social Poetics in the Nation State* (2nd revised edition). New York: Routledge.

Herzfeld, Michael. (2005b) Political optics and the occlusion of intimate knowledge, *American Anthropologist* 107(3): 369–76.

Kapcia, Antoni. (2000) *Cuba: Island of Dreams*. Oxford: Berg.

Lee, Roger. (2006) The ordinary economy: Tangled up in values and geography, *Transactions of the Institute of British Geographers* (N.S.) 31: 413–32.

Long, Norman and Paul Richardson. (1978) Informal sector, petty commodity production and the social relations of small-scale enterprise. In *The New Economic Anthropology*, edited by John Clammer. London: Macmillan, pp. 176–203.

Marston, Sallie A., John Paul Jones III and Keith Woodward. (2005) Human geography without scale, *Transactions of the Institute of British Geographers* (N.S.) 30: 416–32.

Paine, Richard. (1967) What is gossip about? An alternative hypothesis, *Man* (N.S.) 2: 278–85.

Parry, Jonathan and Maurice Bloch. (1989) Introduction: money and the morality of exchange. In *Money and the Morality of Exchange*, edited by Jonathan Parry and Maurice Bloch. Cambridge: Cambridge University Press, pp. 1–32.

Pertierra, Anna Cristina. (2007) *Cuba: the struggle for consumption*, doctoral thesis, University College London.

Roitman, Janet. (2005) *Fiscal Disobedience: an Anthropology of Economic Regulation in Central Africa*. Princeton: Princeton University Press.

Roland, L. Kaifa. (2006) Tourism and the *negrificatión* of Cuban identity, *Transforming Anthropology* 14(2): 151–62.

Roman, Peter. (2003) *People's Power: Cuba's Experience with Representative Government*. Lanham: Rowman and Littlefield.

Sahlins, Marshall. (2004 [1974]) *Stone Age Economics*. London: Routledge.

Sayer, Andrew. (2011) *Why Things Matter to People: Social Science, Values and Ethical Life*. Cambridge: Cambridge University Press.

Scott, James. (1998) *Seeing Like the State: How Certain Schemes to Improve the Human Condition Have Failed*. New Haven: Yale University Press.

Shirk, Susan. (1984) The decline of virtuocracy in China. In *Class and Social Stratification in Post-revolutionary China*, edited by James Watson. Cambridge: Cambridge University Press, pp. 56–83.

Smirnov, A.A. (1965) El desarrollo de la psicología soviética. In *Psicologia Soviética: Selección de Artículos Científicos*, edited by Hans Hiebsch. Havana: Editorial Nacional de Cuba, pp. 1–17.

Smith, Neil. (1992) Contours of a spatialized politics: Homeless vehicles and the production of geographical scale, *Social Text* 33: 54–81.

Stockman, Norman. (1992) Market, plan and structured social inequality in China. In *Contesting Markets: Analyses of Ideology, Discourse and Practice*, edited by Roy Dilley. Edinburgh: Edinburgh University Press, pp. 260–76.

Szulc, Tad. (1989 [1986]) *Fidel: a Critical Portrait*. Dunton Green: Coronet Books.

Trouillot, Michel-Rolph. (2001) The anthropology of the state in the age of globalization: close encounters of the deceptive kind, *Current Anthropology* 42(1): 125–37.

Unión de Jóvenes Comunistas. (1989) Estatutos. *UJotaSeis, Congreso de la Juventud Cubana*. Havana: Editorial Política.

Valdés, Orlando. (2003) *Historia de la Reforma Agraria en Cuba*. Havana: Editorial Ciencias.

Wanner, Catherine. (2005) Money, morality and new forms of exchange in post-socialist Ukraine, *Ethnos* 70(4): 514–37.

Wedel, Janine. (1986) *The Private Poland*. Oxford: Facts on File.

Wilson, Marisa. (2012) The moral economy of food provisioning in Cuba, *Food, Culture and Society* 15(2): 277–91.

Wilson, Peter. (1973) *Crab Antics: the Social Anthropology of English-speaking Negro Societies of the Caribbean*. New Haven: Yale University Press.

Wolf, Eric. (1969) *Peasant Wars of the Twentieth Century*. New York: Harper Books.

Chapter Six
The Scalar Politics of Sustainability: Transforming the Small Farming Sector

I have argued that the Cuban national moral economy provides a background value system that underpins ordinary economic geographies in Tuta, including the consumption and exchange of food (and privilege). In this chapter, I follow along these lines, moving from the consumption and exchange of food to its production. By considering the rise of sustainable small farming in Cuba, I also shift the discussion from social or re-distributive justice to a particular form of environmental justice that has emerged in Cuba since the 1990s, which is tied to a political agenda for food, energy and 'technological' sovereignty.

Similar to my analyses of the consumption and exchange of food and privilege in Tuta, here I outline the spatial and material contours of small-scale sustainable food production through the lens of normative justification. In line with the previous chapters, I show how inequalities wrought by recent openings of the agrarian economy to more privatized production are justified in both official and unofficial terms. Inequalities between welfare and market circuits of food and value are also relevant here, though given that I have already dealt with this issue, the analysis that follows focuses more on the privileged status of small farmers than on different categories of consumers.

The chapter is split into four parts. The first addresses the Cuban government's drive for food (and other) sovereignties, and assesses how well it has carried out this project since the late 1980s. The second provides an historical and normative background to the Cuban agroecology movement and underscores the shifting position of small farmers in Cuba vis-à-vis dominant

Everyday Moral Economies: Food, Politics and Scale in Cuba, First Edition. Marisa Wilson.
© 2014 John Wiley & Sons, Ltd. Published 2014 by John Wiley & Sons, Ltd.

ideas of small farming in Marxism-Leninism. In the third part (divided into two subsections), I explain how the recent 'privatization' of the rural land-scape (the distribution of small plots of land in usufruct) is justified and regulated by national institutions (subsection 1), and then provide a case study to show how 'worthy' farmers are designated and rewarded (subsection 2). The final section illustrates how the privileged positioning of small farmers in Cuba is both justified *and* controlled at the scale of everyday life.

As in earlier chapters, I compare dominant understandings and their moral and institutional foundations with other meanings, practices and concerns demonstrated by people located within and between state and market spaces. I argue that practical and conceptual tools of the Cuban agroecology movement, such as the need to contribute to the national 'patrimony' and to work to improve the nation's environmental resources, invoke associations and meanings that – even if contested – remain unifying forces as they become associated with longer-term values that have shaped Cuban society over time. The political potentialities of Cuban alternatives will be further discussed in the final chapter, when I pull all the threads of this book together by relating moral economies of food in Cuba to theories of alterity, provisioning and justice.

Food and other 'sovereignties' in Cuba

Food sovereignty as a concept was introduced by the *Vía Campesina* (Farmer Road) movement in 1996, and has been continuously modified since then in accordance with an open debate across the world between members of the movement (for more on *Vía Campesina*, see Desmarais 2007). Its basic rationale is to make up for the 'democratic deficit' in the standard concept of food security, which, as Raj Patel argues, may be 'entirely compatible with a dictatorship – as long as the dictator provided vouchers for McDonald's and vitamins, a country could be said to be "food secure"' (Patel, http://rajpatel.org/2009/11/02/food-sovereignty-a-brief-introduction, last accessed 15 May 2013). Food sovereignty means that the structure of each food system (national, regional, etc.) should be determined democratically, and that those who actually support food systems – such as farmers, fishermen, cattle ranchers – should have a direct say in *how* a particular place becomes 'food secure'.

The other two related 'sovereignties' that shape the organization of food production in Cuba are perhaps less well known. Miguel Altieri and Fernando Funes-Monzote (2012: 6) define them as such:

> Energy sovereignty is the right for all people to have access to sufficient energy within ecological limits from appropriate sustainable sources for a dignified

life. Technological sovereignty refers to the capacity to achieve food and energy sovereignty by nurturing the environmental services derived from existing agrobiodiversity and using locally available resources.

The political rhetoric of food and other types of sovereignty suggests the creation of democratic units that work as alternatives to the global capitalist food system. But geographers of 'alternative' or 'diverse' economies recognize the impossibility of such bordered communities. As Amartya Sen writes, no community exists entirely in a 'secluded cocoon' (Sen 2009: 130). All alternatives face 'inherent contradictions and material challenges' (Fuller *et al.* 2010: xxvi) from their embeddedness in wider macroeconomic, political and cultural circuits and networks. Thus we have seen that recent political economic changes in Cuba counter attempts to create an enclosed moral economy of food circumscribed by the nation state. As one Tutaño told me: 'We have to import things in CUCs for tourism, but *They* are trying to make it so that all food Cubans eat is from here'.

Transformations of Cuban agriculture bring to the fore contradictions between state and market spaces, but also between two opposing models for agriculture: the agroecological model, which focuses on 'the whole farm system', and the conventional, industrial model, which focuses on 'individual farm components' (Wright 2012). Despite plans for a 'self-sustaining' (*autosuficiente*) (Castro Ruz 2010) domestic food sector through low-input smaller-scale production, the Cuban government tends to promote at least some large-scale industrial farms during better financial times (Altieri and Funes-Monzote 2012: 5). And while Cuban farmers and scientists have created a wealth of genetically diverse crop strains, the government is experimenting with a transgenetic variety of corn (Bt) which may threaten this emerging biodiversity(Altieri and Funes-Monzote 2012: 6).

For our purposes, the most profound irony is that the government's promotion of food sovereignty does not (or has not yet?) overridden the country's status as a net importer of food, as evidenced most dramatically after 2007 when Cuba suffered from severe hurricanes and after the latest global economic crisis, which began in 2008. As former senior director of the US National Security Council's Office of Inter-American Affairs, Richard Feinberg (cited in chapter 2), writes:

> As a result of shortfalls in domestic agricultural production, Cuba must allocate scarce foreign exchange to feed the population: in 2008, Cuba imported $2.3 billion in foodstuffs including rice ($479 million) and beans ($148 million) and other dietary staples such as powdered milk ($234 million), chicken ($166 million), and fruits and vegetables ($219 million). In response, the government has made increasing agricultural production for domestic consumption a priority. (Feinberg 2011: 11)

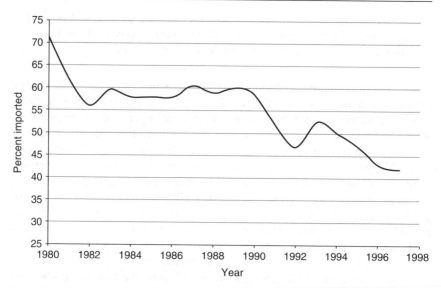

Figure 6.1 Cuba Food Import Dependency, 1980–1997. Source: Álvarez, (2004) Extension Report FE483, http://edis.ifas.ufl.edu/fe483 (last accessed 15 May 2013; from Altieri and Funes-Monzote 2012: 3). Reproduced with permission from José Álvarez and Monthly Review.

The paradox between national food sovereignty and food import dependency becomes less extreme, however, if one considers the relation between food imports and domestic food sources in Cuba. As Figure 6.1 reveals, food imports actually decreased in Cuba from the 1980s, when the Cuban government began to shift towards a model for domestic food production (called the *Programa Alimentario*, Alimentary Programme). Figure 6.1 suggests that Cuba's 'economic battle' for food sovereignty (Castro Ruz 2010) was successful, at least from 1980 to 1998. Writing in 2012, agroecologists Miguel Altieri and Fernando Funes-Monzote claimed that, despite rising levels of food imports after 1998, only 16% were used to satisfy domestic consumption. This means that, in theory at least, Cuban producers were able to cover 84% of domestic alimentary needs (Altieri and Funes-Monzote 2012: 3). 'Overall data show that Cuba's [domestic] food import dependency has been dropping for decades, despite brief upturns due to natural and human-made disasters' (Altieri and Funes-Monzote 2012: 3). Such reports of domestic food production in Cuba are encouraging for those interested in food sovereignty; however, they eclipse inequalities between domestic and tourist/export sectors. Thus as we have seen, some channels for accessing food in Cuba are more adequate than others. The domestic sector that sells food at prices that conform to the purchasing capacity of most Cubans still lags behind more expensive markets that conform to world market prices: 'Here the person with money eats ... There's more food available in CUCs [in 2011, compared to my previous visit in 2007], but no one but tourists can afford it!'

Like all 'diverse economies' (Fuller *et al.* 2010: xxiv), the state-backed programme in Cuba to re-connect domestic food production to the real consumption needs of ordinary Cubans must confront material realities that derive from structural constraints beyond the community's borders. Macroeconomic forces reinforce inequalities between the haves and the have-nots, as do other factors such as the long-term marginalization of small farmers in Cuba. Yet the Marxist-Leninist discourse of small farmers as 'backward' has recently shifted to incorporate at least a minority of dedicated farmers, as detailed in the next section.

Positioning small farmers in Cuba: the agroecology movement in historical context

> The role of power ... becomes most evident in instances where major organizational transformations put signification under challenge. (Wolf 1990: 594)

As in other plantation societies, small-scale producers have been marginalized in Cuban society at least since the time of slavery. For centuries, the organization of land tenure in Cuba has hinged on the large-scale production of monocultures (sugar as well as tobacco and coffee) for export, counterbalanced by a reliance on imported foodstuffs. This plantation pattern continued after the 'triumph' of Fidel Castro's Revolution, especially after Guevara's diversification plan failed by the mid-1960s.

A well-known Cuban author on the first two agrarian reforms in Cuba (of 1961 and 1963) considered small farming in Cuba a 'transitory phenomenon' that would end with the continued development of socialist productive forces (Aranda 1968: 152). In the face of later changes to the agrarian structure, including partial market openings, the Marxist-Leninist hierarchy of property – from state farms at the highest, to cooperatives, and finally, to individual farms that 'will' be eliminated by the voluntary action of small farmers – has been the dominant theory for agrarian planning in Cuba at least since 1963, the year of the second and more radical *Ley de la Reforma Agraria* (Agrarian Reform Law).

Continuing a plantation-like system of monoculture and dependency was not the original intention of the Revolution's architects, however. Indeed, as I have argued elsewhere (Wilson 2010a), Fidel Castro's plans for the agricultural sector shifted from a kind of agrarian populism that emphasized diversification in the pre-1960s (characteristic of other Latin American governments of the period) to a Cuban version of Marxism-Leninism in the early to mid-1960s, under which the large-scale, industrial production of sugar for export continued. The concomitant shift from a more moderate approach to land reform to a radical transformation of the entire agrarian economy required a re-definition of the very category '*campesino*' (farmer), ironically the symbol of revolutionary nationalism in Cuba.

> Despite the heterogeneous demography of the Cuban countryside, there is a long history in Cuba of nationalist outcries against the exploitation of *campesinos*. This history gains moral substance from the idea that nineteenth-century *mambisas* (ex-slave revolutionaries) and *guajiros* (smallholders) were the primary 'war heroes' of the first Cuban revolution of 1868. Indeed, after that time, '*tierra o sangre!* (land or blood!)' was the primary revolutionary call to incite nationalist emotions. (Wilson 2010a: 33)

By contrast, after the adoption of a Marxist-Leninist model, *campesinos* and the newer category of *parceleros* (small plot holders; see below) were to become associated with the lowest form of property ownership in Cuba. But the latter have recently been revived as symbolic exemplars of the kind of revolutionary nationalism that developed in Cuba over a century before Castro's 'triumph' – this time in relation to environmental as well as social justice.

As in Vietnam and China, in Cuba top-down forms of collectivization have dominated the Cuban agriculture sector until quite recently. Yet, as Mark Selden and Benedict Kerkvleit argue for the former two countries (1998: 43), the communist subordination of household farming to more 'social' groupings like the cooperative never entirely eliminated the household economy and its social organization (Selden and Kerkvleit 1998: 43). Indeed, the importance of the household for food production is now officially realized in all three countries, and each has implemented decollectivization drives at various points in recent history.

Talk of decollectivization in Cuba began as early as the 1980s, but it was only in the 1990s that policies were implemented to this end. As opposed to the earlier interest in collectivizing small farmers' lands, Decree Law 640 of September 1994 initiated the re-distribution of small plots[1] of nearly 3 million hectares (about 7.5 million acres) of idle state lands in rural areas as well as over 2600 urban and suburban farms (Altieri and Funes-Monzote 2012: 1) in indefinite usufruct to new smallholders called *parceleros* and their families. More recently, in July 2008, Raul Castro announced Decree Law 259, which allowed for the continued re-distribution of state lands to *parceleros* and to *campesinos* who had held larger plots of land for decades (Hagelberg 2010: 3). The latest policy for agriculture implemented as part of Raul Castro's 2010–2011 economic policies continues the trend towards decentralization, with an explicit goal of 'develop[ing] a food self-sufficiency programme at the municipal level, helping urban and suburban agriculture ... and continuing to encourage low-input production with local resources and animal traction' (*Información* 2011: 32).

Since 1993, over 100 000 Cuban families have petitioned for land, about 40 000 receiving small plots of under a hectare (Altieri 2009: 4). In 2009 (the latest data available), 25% of lands were officially considered 'private', that is, held either privately (by *campesinos*) or in usufruct by *parceleros* (compared to less than 10% prior to the 1990s). Of lands, 42% were held by newly formed cooperatives called Basic Units of Cooperative Production (UBPCs)[2] and older types of cooperatives[3] and the rest (33%) were state or military farms (before the 1990s, nearly 89% of arable land was managed by the state; Altieri 2009: 7).

As opposed to other socialist and post-socialist countries, where decentralization measures led to the full privatization of household farm units, in post-1990s Cuba no land titles have been given out (Enríquez 2000: 20). Moreover, land in Cuba and other (post)socialist societies is 'owned' in a different way than it is in the United States, the UK or in other capitalist countries. Katherine Verdery invented the term 'operational ownership' (Verdery 2003: 55–7) to indicate that property in (post)-socialist societies is not detached from one's obligation to the nation (or community), that is, land is not entirely alienable. In line with socialist philosophies of ownership in other (post)socialist countries, land in Cuba is considered a 'social good that should be owned and profited from in a collective fashion' (Enríquez 2000: 20). Accordingly, property rights in Cuba now entail a moral commitment to the civic responsibility of feeding the nation. Indeed, in the face of post-1990s scarcities of food and energy, values like self-sacrifice and hard work for collective benefits are arguably more important than ever.

Newer *parceleros* more readily adopt this morality than the older class of *campesinos* (Murphy 1999). In post-1990s Cuba, *parceleros* have thus become one primary symbolic link between food and civic responsibility, particularly the obligation to feed all areas of the nation – rural as well as urban – with few external inputs. The rise of urban agriculture in Cuba is just one facet of a larger paradigm shift in Cuba towards agroecological production.

> Much of the spread and success of urban agriculture in Cuba is due to the fact that it is based on local resources and agroecological techniques emphasizing two pillars of agricultural sustainability: integrated pest management and organic soil management. … Garden productivity has been sustained using minimal external inputs, applying principles of agroecology and organic agriculture, which are low cost, environmentally sound, and based on locally available resources. For example the use of chemical fertilizers is prohibited within city limits, and gardeners rely instead on organic fertilizers in the form of chicken or cow manure, compost from household food waste, and, increasingly, worm castings. (Altieri *et al.* 1999: 135)

Cuban authors of the most famous book on the Cuban agroecology movement, *Sustainable Agriculture and Resistance* (Funes *et al.* 2002), claim

Figure 6.2   'Exploited' workers on a large farm take a break for a photo. Source: author.

that the recent shift to agroecological production practices reflects a 'socialist ecology' (Funes *et al.* 2002: 22) that successfully resists pressures from neoliberal nations brought on by the economic crisis of the 1990s. Ideological emphasis on the sustainable production of food for one's household and community also has the practical effect of lessening pressure on the state to supply food (Figueroa Arbelo 1995: 64–5) and high-energy inputs.

Given their important status as food producers, *parceleros* are ideologically and legally separated from the older, larger and more 'egoistic' *campesinos* (Martín Barrios 1984). Though memorialized as the forerunners of 19th- and 20th-century Cuban resistance, from the early 1960s larger farmers were considered a threat to the socialist revolution, 'a detriment to the interests of the working population, who inhibit the production of food for the nation by speculating or using [their farms for] antisocial and counter-revolutionary ends, whose elevated salaries are obtained by exploiting workers' (direct citation from the Second Agrarian Reform Law, Valdés 2003: 179; my translation). Moreover, as Julia Wright (2012) argues, before the 1990s, organic farmers were considered counter-revolutionaries because it was assumed that they produced low yields.

The most obvious legal differentiation between older *campesinos* and *parceleros*, respectively, is that between ownership that may be passed on from generation to generation, and usufruct (or leased) property, which the state may reclaim at any time. Despite this legal difference and in continuation with the Leninist hierarchy of property forms, in Cuban officialdom both *campesinos* and *parceleros* are considered private farmers. Given this status, and similar to other private sectors of society, it is necessary that both types of farmers be 'integrated' into socialist institutions like ANAP (the National Association of Small Farmers), their profits 'controlled' by other state institutions(as illustrated later).

Another legal distinction between *campesinos* and *parceleros* is the require-ment to donate produce for collective consumption. While it is illegal for larger farmers to shun the Acopio (the state institution that re-distributes produce from farmers for 'social consumption'), it is legal for *parceleros* to do so. Given the restricted scale of their productive capacity, *parceleros* are only required to produce for their families and communities, though many also choose to donate to the Acopio (interview with farmer, 2 March 2005). The ideological differentiation between larger private farmers, who 'will be phased out' (Figueroa Arbelo and García de la Torre 1984: 43) and *parceleros*, whose primary civic responsibility is to feed their families and communities, allows Cuban communists to reconcile the recent influx of small farmers with the Leninist idea, still prevalent in Cuba, that some private farmers *may* become a self-interested and privileged class, 'isolated' from the population of workers(Martín Barrios 1984).

Contrary to other farmers who may own private lands and exploit their workers and the rest of the population (Figueroa Arbelo and García de la Torre 1984: 41), *parceleros* are seen as 'social citizens' (Suárez Salazar 2000: 305–13) whose primary interest is to supply their families and com-munities with food. One Cuban author refers to *parceleros* as 'workers for our socialist society who use their plots of land only for the benefit of society, without using more than family members as workers' (Figueroa Arbelo and García de la Torre 1984: 41; my translation). In addition to their positioning as 'social citizens', *parceleros* are also seen as the forerun-ners of the 'socio-agroecological revolution' (*revolución socio-agroecológica*; Suárez Salazar 2000: 292) because of their ability to make agroecology a 'universal science of sustainability' (*ciencia universal de sostenibilidad*; inter-view with agroecologist from the Agrarian University of Havana, 9 January 2005). The new position of *parceleros* even led one Cuban agroecologist to write that Cuba may become the 'first sustainable society of the twenty-first century'(García 2002: 92).

The word 'sustainability' (*sostenibilidad*) and its derivatives are imported symbols in Cuba that were likely adopted after 1990 when Fidel Castro attended the United Nations Earth Summit on Sustainable Development in Rio de Janeiro. Upon his return to Cuba, the former president initiated a shift from the Stalinist model of large-scale, industrial agriculture to smaller-scale production and the low-input model of agroecology. Though there are various definitions of agroecology, including its role as a science, a move-ment and a practice, the standard definition is: 'An ecological approach to agriculture that views agricultural areas as ecosystems and is concerned with the ecological impact of agricultural practices' (http://www.merriam-web ster.com /dictionary/agroecology, last accessed 15 May 2013). As of the year 2000, ANAP has promoted agroecology as their official strategy for small-scale farming (Rodríguez Castellón 2003: 87), particularly through their 'farmer-to-farmer' strategy for the grassroots adoption of agroecological practice. To spread the 'agroecological message to the people' (Funes *et al.* 2002: 15), the communist party also sponsors scientific publications, media coverage, gives funds to national outreach programmes and rewards the most productive farmers with material and moral incentives, most recently, plots of land.

Contrary to the modernist bias of Marxism-Leninism, the importance of traditional agricultural practices in Cuba such as crop rotation, animal traction and the use of family labour is now officially recognized (MINAG 1999). Still, the shift to sustainable, low-input and small-scale farming overturns neither communist norms of collective property nor the principle of scientific Marxism: the idea that technological innovation should be used to further the collective interests of society. For political reasons (e.g. the 'neutrality' of technical language), quantitative over qualitative data collection continues to prevail in extension services to *parceleros*. When in 2005 I conducted research with agroecologists on the 'socio-economic sustainability of *parceleros*' – a title I was encouraged to adopt – I was told that the qualitative data collected on the perspectives of individual farmers were 'not scientific enough'. Scientific Marxism is a key aspect of the Cuban agroecology movement, for without the presence of 'experts' on the farm, *parceleros* risk becoming 'isolated' (Martín Barrios 1984: 91–2) and possibly estranged from social interests. I will return to this aspect of the agroecology movement later in the context of *parceleros*' own experiences of practical and ideological controls over their production.

In defending and promoting the Cuban agroecology movement, Cubans have asserted that they have a right to define and defend their *own* model for agriculture and for the Cuban political economy more generally (Figueroa Arbelo and García Baez 1995; Rodríguez Castellón 1999; Funes *et al.* 2002). Indeed, the so-called '*Greening of the Revolution*' (Rosset and Benjamin 1994) was introduced by the Cuban government to maintain the island's socialist model for economic development (Rodríguez Castellón 1999). The very idea of low-input agriculture was promoted by the Castro brothers in the early 1990s as a way to subvert the kind of neoliberal 'transition' predicted to occur in Cuba after the fall of the Soviet bloc. Similar to the 'victory gardens' in post-war UK, sustainable food production became the highest priority for land use in Cuba (Murphy 1999: 1).

In a society dominated by militaristic propaganda, Fidel Castro's call in the early 1990s for Cubans to start thinking in terms of 'beans' rather than 'bullets' (Eckstein 1994: 28) and 'cannons' (Hagelberg 2010: 1) had a certain appeal to older defenders of the Revolution, many of whom petitioned for plots of lands after 1994. Raul Castro has carried on Fidel's martial slant to promote the agroecology movement, which is now associated with wider discourses of achieving food sovereignty and 'sustainable socialism'. Contradicting his earlier modernist stance, Raul now makes symbolic connections between a lower reliance on imports of food and petroleum-based inputs, on the one hand, and the need to defend Cuba's national borders, on the other. As he stated in 1994: 'In the difficult and unchanging geopolitical and geoeconomic conditions of Cuba, food security is an important way to preserve our sovereignty and national security' (cited and paraphrased by Valdés 2003: 399; my translation). As I will indicate, a similar use of moral or symbolic language is used by *particulares* whom I interviewed in Tuta and elsewhere in Cuba. Perhaps this is not surprising, for the Cuban agroecology movement is directly tied to the recent increase in the number of small low-input farmers.

Controlling the mercenary, designating the worthy: small farming and national institutions

Regulating small-scale production: the Acopio, Fruta Selecta and Granja Urbana

The produce reaped on newly distributed lands is very valuable to Cubans. Indeed, small farmers not only have access to produce of superior quality than that which the state provides, but also to what Caroline Humphrey calls 'manipulable resources': goods necessary for survival such as food, which give people in socialist and other societies 'room to manoeuvre' in exchange (Humphrey 1998: 305). Since many staple foods must be purchased in Cuban Convertible Pesos (CUCs) according to world market exchange rates, Cubans who have petitioned for land find food more of an incentive than wages (Enríquez 2000: 3). Small farmers in Cuba are thus better off than most Cuban workers (Deere *et al.* 1993), notably different than the precarious position of small-scale agriculturalists in most of the 'developing' world. *Parceleros*' privileged position vis-à-vis other workers is rationalized in both official and unofficial terms, forms of justification further explained in this section and the next.

That small-scale producers in Cuba generally have better standards of living than state workers is unique to Cuba as a socialist country, for land distribution in other socialist societies like China and Vietnam guaranteed subsistence but not higher living standards (Selden and Kerkvleit 1998: 53). But because *parceleros* often earn more than workers with peso wages, their presence must be morally justified with socialism, which is based on the precept that all productive members of society should gain equal benefits through

Figure 6.3 A sign outside an agricultural cooperative which reads: 'Agricultural and dairy producers: More unified, strong and with more solidarity than ever before'. Source: author.

equal representation (according to the tenet of democratic centralism). Like workers (and indeed, larger farmers and *particulares*), *parceleros* must therefore be 'organized' by national institutions. As one promotional pamphlet claimed: 'The sustainable use of natural resources … brings possibilities for the development of communities[,] … the equitable distribution of riches, and raising the quality of life *for all those who are integrated into [Cuban] society* (Remon Ramón 2003: 1; my emphasis and translation).

  With the exception of one farmer, whose recognized position as an exemplar of the agroecology movement allowed him to remain independent, most of my interviewees spent much time talking about their relations (and frustrations) with national institutions such as the Acopio. Indeed, for a majority of producers who must sell a large quota of their produce to the Acopio, this re-distributive institution is seen as having many problems that affect both producers and consumers. The 2010–2011 policies make an explicit reference to transforming the Acopio system to 'eliminate losses, simplify connections between primary production and the final consumer, including the possibility of the producer bringing his own products to the market' (*Información* 2011: 29). But while the Acopio system continues to be emphasized in recent Cuban policies, ANAP president Orlando Lugo Fonte still calls it 'an unresolved topic' (see http:// internalreform.blogspot.com/2011/11/co-ops-to-engage-in-food-distribution. html, last accessed 15 May 2013).

The centralization of storage and transport through the Acopio system is a main obstacle to establishing localized producer–consumer networks in Cuba. In the 2005–2007 period at least, nearly every smallholder in Tuta complained to me about wider problems of the Acopio, including the tendency to damage produce or the lack of transport to pick it up in the first place. They also lamented about insignificant payments that the Acopio makes in pesos. One Tutaño *campesino*, the president of his CCS (Credit and Service Cooperative), told me that many smallholders were frustrated as the cheques they received from the Acopio 'have little value because *They* have no money in their account'. Another told me angrily that, 'One has to spend two days working for the Acopio before one has enough to sell for a day at the market!' The same person said that delays in payments from the Acopio can last up to a year. Many other smallholders interviewed complained that the state has not yet provided enough material incentives to produce for the Acopio, though almost every farmer is required to 'donate' (i.e. sell at a very low price) at least half of his or her production to this national re-distributive institution. From my interviews, however, it seems that many do so gladly.

Another institution that regulates where food is directed is called Fruta Selecta. In the 2005-07 period, the tourist market for agricultural and dairy producers was organized through this state enterprise, which processes and sells cleaned and sorted produce to hotels in Havana and other tourist areas, paying most[4] small farmers in an alternative currency (see below). For smallholders, the tourist market regulated by Fruta Selecta or through direct sales (as they have been permitted to do more recently)[5] crosses over into the hard currency economy; it thus allows them to reap more benefits from their production than other members of Cuban society.

To create a buffer in the rural economy between productive earnings in CUCs and the remuneration that most Cubans receive, the Cuban state has established a system of alternative currency to pay *campesinos* and *parceleros* who sell high-quality produce in the tourist market. As with the local exchange trading (LET) schemes in the UK and the Ithaca Hours scheme in New York (alternative currencies that are also tied to non-monetary values created by particular communities), the script money that Cuban smallholders receive from the tourist market becomes a kind of 'special purpose money' (Bohannan and Dalton 1962: 16), which may be used in state stores selling particular goods, such as bicycles, while excluding the sale of others, such as alcohol.

By regulating what is and is not considered acceptable (and available) remuneration, the state sets the standards of value for the way local agrarian economies may enter into the tourist economy. And, as Louis Dumont has argued: 'When the rate of exchange is seen as linked to the basic value(s) of the society', it becomes a 'total social fact' rather than merely an 'economic fact' (Dumont 1986 [1983]: 259–60). As with other alternative currencies, script received in payment for produce destined for the tourist market works as both a 'med[ium] of exchange in the commercial realm and a means for building social relationships in the communal sphere' (Gudeman 2008: 139). Indeed, the script system for remuneration from the Fruta Selecta enterprise has the side effect of encouraging transactions in informal market spaces. As one interviewee who sold bigger, better quality plantains for the tourist sector than those 'that the people eat' said: 'The payments for Fruta Selecta can buy a bike, but they cannot buy clothes. So if the *guajiro* [farmer] needs clothes, he buys a bike, sells it and then buys clothes'. Once again we see that state and market, the formal and the informal collide in ordinary economic spaces.

Until recently,[6] access to the legal tourist market could only be met through the intermediary of Fruta Selecta, which sold farmers' produce to the tourist sector for 'the benefit of all the population'. Like sales made by *particulares*, highly taxed so as to secure some measure of collective responsibility, profits from produce sold in the tourist and other markets are officially justified if they are partially redirected to ensure collective entitlements. An article published in *El Habanero* newspaper on 19 January 2007 outlined the progress of a new, government-funded project to increase fruit production in Havana Province. As an afterthought, the author added: 'For the moment all harvests are for tourism, which benefits the whole population' (Leonardo 2007: 2). However, it was clear to me that some of the profits from the tourist trade were distributed to a few fortunate farmers with access to this market. Farmers' evaded the sensitive issue of differential earnings by always answering my questions with the same comment: 'Cubans don't think about savings'.

Despite the 'humble' dimensions of their economies (as many described themselves), small-scale producers and the intermediaries who may assist in marketing their produce have historically been targeted by the Cuban government as perpetuating the 'evils' of market society. Indeed, while it is recognized that market sales by private producers may temporarily coexist with the centralized, 'egalitarian' distributive system (the Acopio enterprise), the Cuban government and its 'interstitial, supplementary and parallel structures' (Wolf 2001: 167–8) control the risks of profit hoarding and exploitation by petty commodity producers and intermediaries. Otherwise, as Caroline Humphrey argues in her study of consumption in post-socialist Moscow, 'people will resort to [the] dark and violent self-interest' (Humphrey 1995: 44–5) associated with market transactions.

Recent decentralization measures have been coupled with a reinforcement of existing national institutions in the agricultural sector such as ANAP, as well as the creation of new institutions to assist and 'control' *parceleros* such as Granja Urbana. Granja Urbana (translated as Urban Farm Group) is a state institution dedicated to the national programme for urban agriculture and agroecology. It is comprised of extension workers who train farmers in agroecological methods whilst keeping an eye on their income levels and the 'social' aspects of their production (interview with agroecologist from the Centre for Agrarian and Rural Development Education, CEDAR, 11 December 2004). One *parcelero*, who introduced himself as a 'sustainable producer for the Revolution', claimed that the aims of Granja Urbana changed since its inception in the mid-1990s, a shift that has compromised what he claimed to be the overall goal of the agroecology movement: 'To promote creativity and resourcefulness at the farm level so as to increase sustainable production for the local and national community'. According to this farmer, Granja Urbana changed its 'social objectives' over time from financial and technological services (i.e. providing bio-inputs and facilitating the on-farm production of worm humus) to 'control' overproductive techniques. He told the story of a time when he introduced a simple irrigation system on his farm, watering his tomato seeds via an underground tube with holes corresponding to the place where each seed was planted. Upon noticing this act of private initiative, the Granja Urbana inspectors told him that he was not allowed to have this irrigation system as he did not seek permission from the state.

As this story illustrates, state control over production extends to agricultural methods, as it has done at least since the height of agrarian modernism in Cuba. According to the former secretary of the municipal level of ANAP, all smallholders in Cuba have to be organized as members of national institutions which 'orient' and 'control' their production techniques as well as their yields. *Parceleros* and *campesinos* must integrate into their municipal ANAP and attend meetings held by municipal officials of Granja Urbana. When I asked what was discussed and/or accomplished at these meetings, one elderly *campesino* in Tuta complained:

> [They discuss] nothing. Plans, working harder ... that they *will* give us petroleum, fertilizer, machines ... and they never give us anything. ... the Granja Urbana meeting last year lasted five hours! And it took three hours to travel there and back. One day's work lost. ... And they also do not talk about anything – plans, what they *are going* to give us, what they *will do* for us ... and they never do anything.

When I asked an elderly *campesino* in Tuta whether or not he had learned a great deal from publications by *Ciencia y Capacitación* (Science and Learning), a subprogramme of Granja Urbana, he replied, 'what do they have to teach me that I do not already know?' During my several visits to his farm (and many times passing it), I observed a significant amount of pesticide, fertilizer and herbicide use. As exemplified by this *campesino* and as

Julia Wright notes (Wright 2005: 74), many Cubans trained in agriculture prior to the late 1980s still abide by the Stalinist model of high-input industrialized production. This situation has been addressed by the Cuban government, which conduct seminars in agroecology for older farmers and farm managers (*jefes*).

By contrast, *parceleros* usually lack prior knowledge of agriculture and so may be more willing to adopt agroecological methods than the older *campesino* sector. As other researchers have noted, *parceleros* express a high level of commitment to sustainable production: 'Producers spoke passionately and eloquently about their philosophical belief in the need to preserve a balance in the agroecosystem' (Nelson *et al.* n.d.: 5). Catherine Murphy, a sociologist and documentary film-maker of Cuban origin, has argued that *parceleros* as well as officials see themselves as integral parts of the moral community that defines how sustainable agriculture is conceptualised and practised in Cuba: 'Th[e] commitment to share the food harvest is a powerful testament to the spirit of collectivity and solidarity of the Cuban people, and has allowed them to survive the worst moments of the economic crisis' (Murphy 1999: 16).

Given the risks of profit hoarding, however, some *parceleros* are considered more worthy of state benefits than others. Cubans who have received lands in usufruct during the most recent agrarian reform are usually older revolutionaries who are considered exemplars of the 'daily heroism' (*heroismo cuotidiano*) of the post-1990s period of austerity, especially in their tremendous work effort transforming non-productive, damaged lands to productive, sustainable farms. Moreover, according to my experience, only dedicated revolutionaries are eligible to receive land as *parceleros*. Many of the *parceleros* I interviewed had previously 'served Fidel' in Angola in the late 1970s and 1980s. As observed in other socialist contexts (e.g. Pine 1995: 54), most of those who have obtained lands in usufruct have been officials of the Party and/or the Revolutionary Armed Forces (FAR). Access to land during the past two decades is thus linked to what Caroline Humphrey has called the 'transactional value of work' in socialist society (Humphrey 1998: 306): rewards granted from the state for evident work effort for the nation. Below I provide a case study of how *parceleros* localize such discourses in their own words and practices.

Separating the worthy from the unworthy: a case study of Eduardo

As referred to elsewhere (Wilson 2010b; see also Wilson 2012), one of the most dedicated *parceleros* I interviewed had worked as a *jefe* in construction and volunteered during the construction brigades of the 1970s and 1980s. In fact, Eduardo had built many of the residences in which agricultural scientists from the nearby university now live, which made him morally eligible to receive land and, later, *apoyo* (support) from the state-run agricultural university, such as access to information and bio-inputs. In Cuba as in other (post)socialist countries, volunteer work such as that performed by Eduardo has this kind of 'transactional value' (Humphrey 1998: 306), which opens up opportunities for land acquisition and other benefits allotted by national institutions.

Besides engaging in volunteer work, acting as a positive example to represent the Revolution to outsiders is another way to gain access to resources for production. Eduardo and his family seemed to be *the* shining example of the successes of the agroecology movement, and his farm was shown to many visitors (academic or otherwise) of the agrarian university, including myself (first as a researcher for the agrarian university in 2004 and 2005, then as a visitor in 2011. Upon my return to his farm in 2011, I noticed he had added a wind turbine and solar panels). Eduardo was gratified to have completed 27 out of 28 subprogrammes of the urban agriculture programme (later, the number of programmes rose to 29), which are listed in appendix 6. He proudly told me that he had reached the 'excellent' category assigned by Granja Urbana officials, an important asset in preventing 'Them' from taking his farm away.

Eduardo, who lives on the one-third hectare farm with his retired wife, son and daughter-in-law, is seen as an example for the agroecology movement, and is famous at least in the Havana Province for spreading knowledge about agroecological farming. In his own words: 'I am known all over Cuba as someone one can call to get knowledge and help for their planting. … This [farm] is a place of learning'. Eduardo's production and attitude represent the techno-scientific goals of the agroecology movement because he actively supports its 'knowledge-intensive' (rather than 'capital-intensive'; Marrero and Cruz n.d.: 1) aims. His farming methods incorporate traditional knowledge, for he uses plants considered medicinal in Cuba such as saffron, which he claims to be good for the liver. Eduardo also inter-crops plants of different shades of green, which he said 'confuses' insects, and he intercrops corn with mint, the smell of which he said wards off pests. He claimed that knowledge of plants is an essential element of life in Cuba: 'The poor family goes directly to the plants'. Though his household obviously earned more income than most I visited, like other small farmers, he still considered his family 'humble'.

In addition to traditional knowledge, Eduardo incorporates knowledge acquired from visits from agroecologists and other agricultural scientists, such as *in situ* seed and sapling production and the development of a large vat in which he prepares worm humus for application on his crops. He proudly told me that he grows over 105 varieties of plants, entirely with the 'work and sweat of his family's labour'. From what my inexpert eye could recognize, he and his family grows coffee (though he emphasized that this is 'only for family consumption' – coffee sales are monopolized by the state), pumpkins, cassava, taro, string beans, plantains, bananas, cocoa, avocado saplings, cherries, lemons, mint, oregano, rosemary, coriander, basil, garlic, tomatoes, *guayabas* (a type of guava), figs and ornamental plants. He also raises catfish, rabbits, chickens and a goat (notably, he did not raise pigs as many other farmers do 'on the side'. This is likely because he is frequently visited by Granja Urbana and other officials). When I visited his farm, Eduardo's primary economic activities were selling avocado saplings to visitors and mint to 'people from Havana' who then acted (mostly) as illegal intermediaries, selling the mint to hotels 'for the tourists' *mojitos*[7] and paying the farmer in CUCs. Eduardo also sold guava, tubers and vegetables to visitors in pesos. He told me that he preferred to grow and sell many types of crops since monocultures are 'too risky'.[8]

Eduardo's position as 'excellent' in the Granja Urbana tier system is clearly an important element not only in accessing inputs and knowledge, as well as retaining his land, but also in moral terms, for he and his family expressed a sense of pride and prestige in representing the Cuban agroecology movement. Besides official recognition as 'excellent', another of the symbols which seemed to represent the prestige of national recognition was the visitors' notebook, in which foreign students and other visitors make comments and compliments about farms they visit. When I became accustomed to the ritual of notebook signing, I routinely asked *parceleros* to see their notebooks. One *parcelero* seemed ashamed to tell me that he did not yet have one: 'I am going to get a visitors' notebook. That way the people at the university and other officials can see that my farm is good'.

But Eduardo proudly showed me his visitors' notebook as did many other interviewees. To Eduardo and other *parceleros*, a visitors' book full of positive comments such as: He is 'striving to be first' in agroecology, or 'He has increased production of guavas and can now lower his prices' – are non-market incentives that are taken seriously by many producers. Such moral incentives not only affect methods of production and output, but also work to enable *campesinos* and, especially, *parceleros* to become (worthy) persons in Cuban society.

Perhaps the best way to understand the importance of acting as an example of the Revolution for outsiders through agroecological production is when this rule is broken. One *parcelero* did break the rule, which had adverse consequences both for himself and this author. The farmer was a man in his late fifties who one day exposed himself to me lasciviously. The professors at the university in charge of my 'protection' had strongly encouraged me to come to them with any problems during my stay. I eventually, though reluctantly, told them about the incident. After this, all *apoyo* (support, both material and moral) that the university and other national institutions had provided to the farmer was cut off. I would not be surprised if he even lost the land on which he worked as a *parcelero*. As a professor told me: 'We cannot let this man ruin the image of the Revolution with people like you'.

His comment recalls the idea of 'protecting' tourists from outsiders within, an issue addressed in the previous chapter. Such stories not only reveal how important it is for people in Cuban society (like smallholders) to be 'in line' with national institutions, but also remind us of how Cuba's scalar politics affects relations with people from outside national borders, like myself.

Shifting borders of Cuban agroecology

As with other food movements, the normative delineation of Cuban agroecology's boundaries is often contested from within, even if challengers use similar nationalist representations to those created and preserved by the ideological structure that drives them to question the status quo in the first place. Thus while Cubans involved in the agroecology movement may defend the official idea that citizens recently given land in usufruct should dedicate themselves to sustainable production for the collective needs of their families and for the Cuban people as a whole, this does not mean that the boundary circumscribing *their* agroecology movement occupies the same conceptual and practical space as that imposed from above.

Still, national tropes were an important way farmers characterized their agroecological practice. *Parceleros* I spoke with framed low-input agriculture in terms of 'defending' the nation's biodiversity (Remon Ramón 2003: 1) in the face of 'enemies' who posed a threat to national sovereignty. Indeed, as I discovered through countless conversations and informal interviews with *parceleros*, the need to protect the nation's biological resources against 'enemies' was especially important after a widely publicised story spread in Cuban newspapers and on the ground about the 'new biological North American aggression of the US' (Súarez Salazar 2000: 313–14). This statement is associated with the rumour that US planes released a plague of biological agent *Thrips palmi* on Cuban crops in 1997.

A related attitude towards outsiders (within?) was expressed by one interviewed *parcelero* who contrasted 'money-minded people' who 'care only about selling things to people that they do not need' with his own interests in producing healthy food for the 'patrimony of the nation'(*patrimonia de la nación*):

> In life, not everything is money. I invent things [e.g. a timed irrigation system using materials such as plastic litre bottles] to grow food, not to make money. Other people would want to sell inventions like mine, to patent them to make money. Anyone ... can use my inventions, and I will not ask them for any money. I make them for the patrimony of the nation.

When I asked another *parcelero* why he chose to work harder than other farmers in Cuba who use chemicals, he responded that the land and its products belong 'to the Cuban nation', and that because of this he must

work hard to ensure the 'future wealth of the country'. The farmer cited
the national hero and first Cuban revolutionary, José Martí – 'one has to
make the world a better place' (*hay que mejorar el mundo*) – in so doing
linking his ecological production to a more profound notion of progress
and wealth embedded in the history of Cuba as a revolutionary society.
Links between José Martí's ideas, national resources and ecological
production were also evident in some of the literature promoting the
agroecology movement, published by academics from the Agrarian
University of Havana. In arguing that low-input, local agriculture works
for the good of the nation's health and well-being, one pamphlet used
Martí's words – 'Agriculture is the only constant, certain and eternal
source of [the nation's] riches'(Remon Ramón 2003: 2).

Parceleros drew from such nationalist understandings of localised, ecological agriculture, such as the correlation between national wealth and
improving the soil (ironically, one of Adam Smith's requirements for sustainable estate management; Barber 1967: 51–4). Many also framed their
practice in terms of socialist principles, such as the pre-eminence of
social over monetary exchanges. In accordance with the idea of land and
food as a form of shared natural (as opposed to individualised monetary)
wealth, one *parcelero* told the story of when visitors from the United
States (over 50 undergraduate students) came to his farm to ask questions. After offering each of his students an 'organic' *guayaba* as a snack,
one of them asked: 'Why are you giving these guavas away! In the US, you
could sell this amount of guavas for lots of money! You could get over
$500 [USD, about 12000 pesos, more than the average salary for 3 years]
for the amount of guavas you have here!' The *parcelero* telling me the story
seemed proud of his response: 'I am not interested in money. I am interested in friendship. I am interested in showing the people how to produce
on the land in a sustainable fashion'.

Exchanges made both on and off the farm reflected his cultural and political conviction that sharing is more important than profit. One day I noticed
that the *parcelero* provided free information for the woman who worked in
the *tienda de la agricultura* (agricultural store), where seeds as well as farming implements could be purchased in pesos. The woman was keen to access
techno-scientific knowledge about agroecology that the farmer obtained
from state institutions, which would likely become a kind of 'manipulable
resource' (Humphrey 1998: 305) with which she would be able to manoeuvre in official settings. In exchange, the farmer sought market information,
that is, clients who may have been interested in his saplings, people selling
seeds that he might have been interested in purchasing, and so on.

Social exchanges such as these, which again reflect spaces in between
state and market, were not always legal. I met *parceleros* who preferred to
exchange crops for manure from neighbouring cattle ranches instead of purchasing pre-made compost legally from official sources (e.g. the university).

Figure 6.4   The inside of a makeshift greenhouse on Eduardo's farm, where avocado saplings are grown organically. Source: author.

One *parcelero* chose to offer rum to informal workers who were then 'socially obliged to continue to work on the farm' instead of hiring state 'specialists' to clean his pig pen. One reason he may have done this is because he was rearing an illegal number of sows and wanted to hide this fact from state officials. Ironically, his preference for social exchanges contradicted the state's requirement to use money to hire state workers.

In addition to legal controls over who one may hire to complete farm tasks, the Cuban government has also increased the required number of visits made by officials and scientists to check farmers' production practices and incomes. Indeed, despite recent campaigns such as *autogestión participativa* (participative management; see below), we have seen that scientific 'experts' and officials still have a powerful place in production. Many of the *parceleros* interviewed felt an estrangement from agroecologists and other extension officers who came to their farms to instruct them in the use of biological inputs such as Ecomic.[9] Eduardo, for example, complained that scientists 'do not listen to me when I explain my needs'. Another *parcelero* told me he preferred to get information and inputs by exchanging information and goods with people in their community: 'I always prefer social relations over science. Scientists say one thing and do another'.

According to some *parceleros* at least, the campaign for *autogestión participativa* – the horizontal spread of agroecological knowledge from farmer to farmer rather than the vertical organization of information dissemination – has not yet proved successful. While Cuban authors (e.g. Suárez Salazar 2000: 391–2) have suggested that the concept of *autogestión participativa* means a greater degree of civil society participation in the agriculture sector, the campaign has also been accompanied by an increase in farm visits from agricultural 'experts'. Both internal and external critics have argued that shifts to more ecological forms of production in Cuba are stifled by the politics of science. According to one author, a complete ecological shift would 'require that the socialists abandon habits and modes of thought long established [in socialist Cuba]; especially the supposition of scientific authority'(Benton 1999: 184; my translation).

Parceleros did not always consider official parameters of the agroecology movement to be the most preferable means to achieve its goals. Careful to point out that they were not interested in profits (or 'savings'), many justified black market activity such as exchanging bio-inputs, information and other 'manipulable resources' in terms of finding the best way to serve the population of consumers. For instance, one *parcelero* complained that the types of seeds the state provides for farmers through state stores and through the institution of Granja Urbana do not produce high-quality produce demanded by the population. He complained that Granja Urbana distributes lesser-quality seeds to *parceleros* who produce for the population, while better quality seeds are reserved for production for the tourist or export markets.

> The seeds that Granja Urbana provides us [small farmers] do not resolve anything. The people do not want green tomatoes or small, tasteless tomatoes; they do not care whether they are produced without chemical inputs. These seeds do not respond to consumer needs. ... They [the government] care more about our image as a sustainable society [*sociedad sostenible*] to outsiders than what the consumer in Cuba really wants. They care more about politics than food!

As illustrated by this comment and other cases presented above, there is not always a consensus about how national understandings of agroecology as a movement (e.g. the civic responsibility to produce sustainably for the nation) may align with more personal concerns like how to define consumer 'needs'. As I have already argued (see Wilson 2009), there is a growing contradiction in Cuba between the official view of food as an entitlement for all needy citizens, and the market treatment of food as a commodity: 'People do not buy [these] green tomatoes! [The state] produce[s] them to avoid plagues, but no one eats [them]! ... In the black market, I can buy foreign seeds that produce large, red tomatoes. Consumers want this kind of tomato!' The dilemma this farmer faces reflects competing politics of scale: the one connected to the socialist nation state, the other to the household or individual. The latter is, in turn, connected to *inter*national market networks, in this case the market for higher-quality tomato seeds.

Agroecology and other projects for sustainability are multiple and shifting, but there remains a hierarchy of dominant meanings and practices effected *and effective* at particular scales. Here I have provided ethnographic detail of farmers' words and experiences which complicate official and/or scientific meanings of agroecology as a national movement in Cuba. Earlier I outlined Leninist hierarchies of ownership, which in Cuba are linked to national(ist) values of sovereignty, work, civic responsibility and national patrimonies, and on the principle of scientific Marxism, which leads to the on-farm presence of agroecologists (and other 'experts') as social and ecological watchdogs. I compared national understandings and their moral underpinnings, some of which were shared by my interviewees, to other meanings, actions and concerns demonstrated by farmers working within or for Cuba's political drive for food sovereignty. As this discussion reveals, we cannot take simplified models of Cuban food sovereignty or agroecology at face value, for there are underlying tensions between official projects for sustainable cultivation projected at the scale of the nation, and the practices and needs of smaller Cuban farmers (and consumers) at the everyday level.

Still, Cuban agroecology may be a more 'alternative' alternative to the global capitalist food system than related but more localized movements like Community Supported Agriculture, or to ethical consumer movements that reach global proportions, such as Fair Trade. Ideological formulations of small-scale sustainable agriculture in the UK or United States, for example, are linked to different political and moral economic histories than to those that have developed in Cuba. 'Going green' in the UK or the United States is, at least for some, as much about consumer freedom as it is about saving the planet. Through a series of events and ideological associations, a 'symbolic convergence' (Needham 1975) has formed relating sustainable food production to other Euro-American values like individualized property, identity as linked to consumption or the fulfilment of personal desires, and a nationalist sense of environmental stewardship (e.g. Aldo Leopold's idea of 'wilderness' preservation in the United States, Harvey 1996: 168–71). Norms used to define and circumscribe alternative food movements acquire tacit meanings over time, understandings embedded in particular social, spatial and material contexts.

Although Cuban agroecology shares with non-Cuban alternatives similar political and moral origins as a counter to mainstream industrial production, it is uniquely emplaced in relation to the nation state. Like all other 'counter-cyclic' (Polanyi 1944) models, however, this alternative programme is characterized by a scalar politics that includes some voices and excludes others. Indeed, while some may extol the merits of Cuba's 'sustainable socialism', others may see it as a kind of militant particularism[10] that favours some knowledges and practices over others. As with all moral economies – including the dominant moral economy of neoliberalism – alternative food networks may become what I have called Leviathans if a singular perspective of 'the good' drowns out other 'modes of ordering' (Law 1994, cited by Whatmore and Thorne 1997: 240) social and material space.

Members of the Cuban agroecology movement delineate their *own* boundaries of the movement, limits that do not always coincide with nationalist meanings and practices. Many of their implicit understandings, however, are based on particular referents and combinations that stem from shared moral, ideological and political economic frameworks. Deeply embedded values, such as the idea of land as a form of collective patrimony, give individual voices in the movement a sense of wholeness, a fuzzy shape that may resemble unity more clearly to outsiders than to insiders.

Notes

1 According to the law, *parceleros* must hold no more than 0.25 ha (or about 0.6 acres) of land.
2 Basic Units of Cooperative Production (Unidades Básicas de Producción Cooperativa) were formed in 1994 by consolidating state lands to remedy the 'giganticism' and low-productivity of the state farm sector (though some still see UBPCs as too large for agroecological production (Figueroa Arbelo 1995: 65; Altieri 2009)).
3 Besides the newer UBPCs, there are two other types of cooperatives in the Cuban agriculture sector. The virtually obsolete Agricultural Production Cooperatives (CPAs) are based on lands pooled by small farmers in the 1970s to be worked collectively, while in Credit and Service Cooperatives (CCSs) lands remain separate though 'service' (e.g. marketing) and credit are managed collectively.
4 Julia Wright (2009: 142) claims that farmers of export crops such as sugarcane and tobacco (who are not members of cooperatives) are paid in hard currency.
5 As of 1 December 2011, small farmers are legally permitted to sell directly to tourist hotels and other markets, though they still must donate a quota to the Acopio. This is a fascinating shift about which I have yet to conduct ethnographic research. Because I am not yet familiar with how this market opening has affected farmers on the ground, I focus my attention on relations between state institutions and farmers prior to the implementation of this policy.
6 See the previous endnote.
7 A Cuban drink made with rum, sugar, soda water, ice and mint.
8 This strategy for production recalls Michael Lipton's well-known thesis that rather than being profit maximizers, 'peasants' are 'optimizers', balancing profits and yields with risks (Lipton 1968).
9 Ecomic is a bio-fertilizer made of mycorrhizal fungi that allows for more nutrient absorption (Nelson *et al.* n.d.: 8). Since 1990, over 220 small laboratories and production centres have been constructed for the manufacture of biological inputs such as Ecomic. Along with chemical inputs, which are strictly regulated (Nelson *et al.* n.d.: 10), biological inputs manufactured by scientists from the university are distributed to smallholders according to their expressed need and ideological commitment to the movement.

10   Harvey and Williams (1995), Harvey (1996: 32–45). David Harvey has taken this term from Raymond Williams, who originally used it to mean: 'Ideals forged out of the affirmative experience of solidarities in one place [*or time*] get generalized and universalized as a working model of a new form of society that will benefit all of humanity' (Harvey and Williams 1995: 83; my addition and italics).

References

Altieri, Miguel. (2009) La paradoja de la agricultura cubana: Reflexiones agroe-cológicas basadas en una visita reciente a Cuba. http://www.ecoportal.net/content/view/full/87045 (last accessed 16 May 2013).

Altieri, Miguel and Fernando Funes-Monzote. (2012) The paradox of Cuban agriculture, *Monthly Review* 63(8): 1–6.

Altieri, Miguel, Nelso Companioni, Kristina Cañizares, Catherine Murphy, Peter Rosset, Martin Bourque and Clara I. Nicholls. (1999) Greening of the 'barrios': urban agriculture for food security in Cuba, *Agriculture and Human Values* 16: 131–40.

Álvarez, José. (2004) The issue of food security in Cuba (Paper FE483). Institute of Food and Agriculture Sciences (UF/IFAS), University of Florida, http://edis.ifas.ufl.edu/fe483 (last accessed 15 May 2013).

Aranda, Sergio. (1968) *La Revolución Agraria de Cuba*. Mexico: Siglo Veintiuno Editores, S.A.

Barber, William. (1967) *A History of Economic Thought*. Middlesex: Penguin Books.

Benton, Edward T. (1999) El enverdecimiento del socialismo: un nuevo concepto de 'progreso'? In *Cuba Verde: En Busca de un Modelo para la Sustentabilidad en el Siglo XXI*, edited by Jesús Delgado Díaz y Carlos. La Habana: Editorial Jose Martí, pp. 184–96.

Bohannan, Paul and George Dalton. (1962) Introduction. In *Markets in Africa*, edited by Paul Bohannan and George Dalton. Evanston: Northwestern University Press, pp. 1–26.

Castro Ruz, Raúl. (2010) Discurso pronunciado por el general de ejército Raúl Castro Ruz, Presidente de los Consejos de Estado de Ministros, en el quinto period ordinario de sesiones de la VII Legislatura de la Asamblea Nacional del Poder Popular. Palacio de Convenciones, 1 Aug.

Deere, Carmen, Ernel González, Niurka Pérez and Gustavo Rodríguez. (1993) Household incomes in Cuban agriculture: a comparison of the state, cooperative and peasant sectors, *Institute of Social Studies Working Paper Series* No. 143, The Hague, pp. 1–50.

Desmarais, Annette Aurélie. (2007) *La Vía Campesina: Globalization and the Power of Peasants*. London: Pluto Press.

Dumont, Louis. (1986 [1983]) *Essays on Individualism: Modern Ideology in Anthropological Perspective*. Chicago: University of Chicago Press.

Eckstein, Susan. (1994) *Back From the Future: Cuba under Castro*. Princeton: Princeton University Press.

Enríquez, Laura. (2000) Development Report no. 14: Cuba's new agricultural revolution: the transformation of food crop production in contemporary Cuba. Oakland: Food First, pp. 1–14.

Feinberg, Richard E. (2011) *Reaching Out: Cuba's New Economy and the International Response*. Washington, DC: Latin American Initiative at Brookings.

Figueroa Albelo, Victor and Román García Báez. (1995) La reforma económica en Cuba y sus direcciones principales. In *El Sector Mixto en la Reforma Económica Cubana*, edited by Nada Ramón Sánchez and Nelson Labrada Fernández. Havana: Editorial Felix Varela.

Fuller, Duncan, Andrew E.G. Jonas and Roger Lee. (2010) Editorial introduction. In *Interrogating Alterity: Alternative Economic and Political Spaces*, edited by Duncan Fuller, Andrew E.G. Jonas and Roger Lee. London: Ashgate, pp. xxiii–xxx.

Funes, Fernando. (2002) The organic farming movement in Cuba. In *Sustainable Agriculture and Resistance: Transforming Food Production in Cuba*, edited by Fernando Funes, Luís García, Martín Bourque, Nilda Pérez and Peter Rosset. Oakland: Food First, pp. 1–26.

Funes, Fernando, Luís García, Martín Bourque, Nilda Pérez and Peter Rosset (eds). (2002) *Sustainable Agriculture and Resistance: Transforming Food Production in Cuba*. Oakland: Food First.

García, Luís. (2002) Agroecological education and training. In *Sustainable Agriculture and Resistance: Transforming Food Production in Cuba*, edited by Fernando Funes, Luís García, Martín Bourque, Nilda Pérez and Peter Rosset. Oakland: Food First, pp. 90–108.

Gudeman, Stephen. (2008) *Economy's Tension: the Dialectics of Community and Market*. New York: Berghahn Books.

Hagelberg, G.B. (2010) If it were just the *marabú*: Cuba's agriculture 2009–10. Published by the Association for the Cuban Economy, emailed to the author, 30 Aug. 2010, 1–18.

Harvey, David. (1996) *Justice, Nature and the Geography of Difference*. Oxford: Blackwell.

Harvey, David and Raymond Williams (1995) Militant particularism and global ambition: the conceptual politics of place, space, and environment in the work of Raymond Williams, *Social Text* 42: 69–98.

Humphrey, Caroline. (1995) Creating a culture of disillusionment: consumption in Moscow, a chronicle of changing times. In *Worlds Apart: Modernity through the Prism of the Local*, edited by Daniel Miller. London: Routledge, 43–68.

Humphrey, Caroline. (1998) *Marx Went Away but Karl Stayed Behind* (updated edition of *Karl Marx Collective: Economy, Society and Religion in a Siberian Collective Farm*, 1983). Ann Arbor: University of Michigan Press.

Información. (2011) *Información sobre el Resultado del Debate de los Lineamientos de la Política Económica y Social del Partido y la Revolución*. VI Congreso del Partido Comunista de Cuba. Havana, May.

Leonardo, C.A. (2007) Proyecto para aumentar frutales, *El Habanero* 19 Jan., 2–3.

Lipton, Michael. (1968) The theory of the optimizing peasant, *Journal of Development Studies* 4(3): 327–51.

Marrero, Pablo and Orestes Cruz. (n.d.) La influencia de la luna sobre los cultivos, unpublished paper, Agrarian University of Havana, 1–15.

Martín Barrios, Adelfo. (1984) Historia política de los campesinos cubanos. In *Historia política de los Campesinos Latinoamericanos*, Vol. 1, edited by Pablo González Casanova. Mexico: Siglo Veinti Uno Editores, pp. 40–92.

MINAG (Ministerio de la Agricultura, Cuba). (1999) *Lineamentos para los Subprogramas de la Agricultura Urbana para el Año 2000*. Havana: MINAG.

Murphy, Catherine. (1999) *Cultivating Havana: Urban Agriculture and Food Security in the Years of Crisis*. Oakland: Food First.

Needham, Rodney. (1975) Polythetic classification: convergence and consequences, *Man* (N.S.) 110: 349–69.

Nelson, Erin, Steffanie Scott, and Angel Leyva Galán. (n.d.) Institutionalizing sustainable agriculture: Opportunities and challenges in Cuba. http://www.organicagcentre.ca/ResearchDatabase/res_socsci_07_nelson.asp (last accessed 15 May 2013).

Pine, Francis. (1995) Kinship, work and the state in post-socialist rural Poland, *Cambridge Anthropology* 18(2): 13–22.

Polanyi, Karl. (1944) *The Great Transformation: the Political and Economic Origins of Our Time*. Boston: Beacon Press.

Remon Ramón, Maydelin. (2003) Por una agricultura sostentible y mayor producción, *El Lajero Andante* (Bulletin No. 2), Año 1, p. 1.

Rodríguez Castellón, Santiago. (1999) La evolución y transformación del sector agropecuario en los noventa (accessed in photocopy form from the Agrarian University of Havana, which did not include bibliographic information).

Rodríguez Castellón, Santiago. (2003) La agricultura urbana y la producción de alimentos: la experiencia de Cuba, *Cuba Siglo XXI* 30: 77–101.

Rosset, Peter and Medea Benjamin. (1994) *The Greening of the Revolution: Cuba's Experiment with Organic Agriculture*. Melbourne: Ocean Press Distributors.

Selden, Mark and Benedict Kerkvleit. (1998) Agrarian transformations in China and Vietnam, *The China Journal* 40: 37–58.

Sen, Amartya. (2009) *The Idea of Justice*. London: Penguin Books.

Valdés, Orlando. (2003) *Historia de la Reforma Agraria en Cuba*. Havana: Editorial Ciencias.

Suárez Salazar, Luis. (2000) *El Siglo XXI: Posibilidades y Desafíos para la Revolución Cubana*. Havana: Editorial de Ciencias Sociales.

Verdery, Katherine. (2003) *The Vanishing Hectare: Property and Value in Post-socialist Transylvania*. Ithaca: Cornell University Press.

Whatmore, Sarah and Lorraine Thorne. (1997) Nourishing networks: alternative geographies of food. In *Globalising Food: Agrarian Questions and Global Restructuring*, edited by David Goodman and Michael Watts. London: Routledge, pp. 287–304.

Wilson, Marisa. (2009) Food as a good versus food as a commodity: contradictions between state and market in Tuta, Cuba, *Journal of the Anthropological Society of Oxford* (N.S.) 1(1): 25–51.

Wilson, Marisa. (2010a) The revolutionary revalorization of campesinos (peasants) in Cuban history and implications for food security in present-day Cuba, *History in Action* 1(1): 1–6.

Wilson, Marisa. (2010b) Embedding social capital in place and community: towards a new paradigm for the Caribbean food system. In *World Sustainable Development*

Outlook, 2010, edited by Gale T. Rigobert and Alam Assad. Sussex: World Sustainable Development Publications.

Wilson, Marisa. (2012) The moral economy of food provisioning in Cuba, *Food, Culture and Society* 15(2): 277–91.

Wolf, Eric. (1990) Distinguished lecture: facing power, *American Anthropologist* 92(3): 586–96.

Wolf, Eric. (2001) *Pathways of Power: Building an Anthropology of the Modern World,* with Sydel Silverman. Berkeley: University of California Press.

Wright, Julia. (2005) *Falta petroleo. Perspectives on the emergence of a more ecological farming and food system in post-crisis Cuba,* doctoral thesis, Wageningen University.

Wright, Julia. (2009) *Sustainable Agriculture and Food Security in an Era of Oil Scarcity: Lessons from Cuba.* London: Earthscan.

Wright, Julia. (2012) Mainstreaming agroecology, Presentation for the Annual Meeting of the Royal Geographical Society–Institute of British Geographers, Edinburgh, 5 July.

Chapter Seven
Conclusion

[E]conomics ... has not really succeeded in emancipating itself from morality.
(Dumont 1977: 172)

In this book, I have treated the concept of moral economy as does Andrew Sayer (2000: 80): 'as both an object of study and a kind of inquiry'. As an *object of study*, I have considered the concept in terms of the idea of economy, which has diverged into two contradictory though complementary moral and political economic projects: welfare socialism and liberal capitalism. This binary is a cultural relation, an 'artefact' of western intellectual history that shapes our ideas and practices of economy:

> Any coherence that exists about the idea of economy derives essentially from our cultural beliefs, which (in Anglo cultures at least) have led to constructions of economy being overlain with the dichotomy of planned versus market, which, in turn, has had the effect of denying the existence of multiple forms of economy. (O'Neill 1997, cited by Williams 2005: 69)

Cuba is a laboratory for the kind of research that seeks to map out multiple and shifting economic logics and practices in everyday life. On a more formal level, socialism and liberalism coexist in Cuba as competing 'social relations of value' with different material bases and spatialities (Lee 2011b: 198). To concretize these moral economic projects as 'objects of study', my focus has been on how they shape two vital pathways of food provisioning

Everyday Moral Economies: Food, Politics and Scale in Cuba, First Edition. Marisa Wilson.
© 2014 John Wiley & Sons, Ltd. Published 2014 by John Wiley & Sons, Ltd.

in Cuba: one attached to the planned national economy, the other to 'global' neoliberal markets. Such an analysis aligns with geographical studies that relate food markets to their moralities and to shifting 'narratives of geographical scale and historical temporality' (Jackson *et al.* 2008: 21). It also contributes to studies of alternative forms of economy, as I explain later.

Along with viewing the concept of moral economy as an object of study, which, in the Cuban case at least, illuminates two competing normative projects for community, each with different materialities and 'space-times' (Castree 2009), I have also treated the concept of moral economy as a *kind of inquiry*. Methodologically, I have considered everyday 'micro' negotiations of economic life in the light of 'macro' projects for moral economic communities, or what I have called Leviathans. Such a relational view of discourse and performativity has helped me to avoid what Michel Callon and Bruno Latour call the 'two fallacies' of social science, that is, either

> giving the macro-actor too much credit or assuming that individual micro-negotiations are truer or more real than the abstract, distant structures of the macro-actors. Here again, nothing could be further from the truth for almost every resource is utilized in the huge task of structuring macro-actors. Only a residue is left for the individuals. What the [social scientist] too hastily studies is the diminished, anaemic being, trying hard to occupy the shrinking skin left to it. In a world already structured by macro-actors, nothing could be poorer and more abstract than individual social interaction. (Callon and Latour 1981: 300)

The very concept of 'moral economy' typifies this fundamental dilemma in the social sciences, for multi-scalar projects for *economy* must establish abstract means to evaluate, measure and calculate, while *moral* frameworks emerge in particular places and are reproduced in daily life. Thus my inquiry has necessitated *both* 'armchair' research of the way the Cuban economy has been constructed in the past and present, *and* long-term ethnographic research in the place of Tuta (Figure 7.1). If social life reflects a cyclical process between everyday actions and existing ideas and norms (Giddens 1984), then part of the ethnographer's job is to undertake discursive and historical study of how such ideas – evidenced by the researcher in everyday life – have emerged in the first place.

In this concluding chapter, I relate this two-pronged analysis of moral economies of food in Cuba to wider debates about 'alterity' in economic geography, focusing not only on moral economic *abstractions*, such as Cuban socialism, which I have treated as an alternative to neoliberal projections for Cuban transition, but also on how these relate to moral economic *practices*, as these enter into and out of two separate, dominant projects for economy. The first half of the chapter (divided into two subsections) indicates some contributions my analysis makes to studies of alternative economic geographies,

Figure 7.1   A scene from the back streets of Tuta. Source: author.

particularly alternative food networks (AFNs; the first subsection). Then I draw from the systems of provisioning approach to explain certain potentialities and drawbacks of domestic food provisioning in Cuba (the second subsection). Taking heed of Andrew Sayer's (2000) approach to moral economies, which points to the epistemologies and methodologies described above but also to the need to make normative conclusions, in the final part of the book I broaden my perspective to consider whether and how the Cuban food system may be viewed as a viable alternative that enables social, economic and environmental justice. To do so, I propose a non-parochial idea of justice and value(s) by drawing from Amartya Sen's (1999, 2009) philosophical insights.

The objective of this final chapter is twofold. The first is to return to theoretical issues covered earlier in the book, particularly the relational and scalar dimensions of moral economies vis-à-vis everyday moral economic life, and to connect this discussion with wider debates in the literature on alternative economic geographies and systems of provisioning. The second aim of the chapter – and the underlying aim of this book – is to offer a normative perspective on justice and value(s) by highlighting multiple positionalities and conceptions of the 'good life' as defined in both non-market *and* market terms.[1] As I argue, whether alternative projects for food provisioning truly broaden the possibilities for change depends on how successful they are at 'jumping scale' from place-based forms of militant particularism to heterogeneous values and ideas of justice, some of which are justifiably embedded in market relations.

Alternative economic geographies and systems of provisioning: contributions and possibilities

The domestic food economy in Cuba as an alternative economic geography

In his preface to Paul Bohannan and George Dalton's classic work in economic anthropology, *Markets in Africa* (1962), Melville J. Herskovits argued against the kind of analysis that poses the capitalist market as a homogeneous social fact that applies to all societies in the same way (see also Miller 1997). Like more recent work in geography (e.g. Gibson-Graham 1996, 2006; Leyshon *et al.* 2003; Lee 2006, 2011a, 2011b; Fuller *et al.* 2010), Herskovits and other anthropologists of the substantivist school claim that societies with combinations of different economic logics such as reciprocity and what they called 'money-barter' (or alternative currencies) are different *in kind* (not just degree) from societies in which market prices determined most formal economic activity (Bohannan and Dalton 1962: 16). Similarly, Roger Lee has argued that 'alternative-oppositional' (Fuller and Jonas 2003: 67) economic geographies – those which work in active resistance against a capitalist mainstream – are different *in kind* from those located nearer the mainstream end of the alternative continuum, for they work according to 'something different in value or operational terms'(Lee 2010: 280).

Throughout this book, I refer to the Cuban moral economy as an alternative to the moral economy of neoliberalism, but I have not yet explained in detail what I mean by 'alternative'. Following geographers of alternative economic spaces, I define 'alternative' as 'spaces of social commons against the enclosing logic of capital' (Hodkinson 2010: 242). In this light, the domestic food economy in Cuba may be seen as a distinct sphere with its own social relations of value, power geometries and materialities. As I have argued, the national food basket in Cuba – increasingly associated with the 'needy' – is a public good, a right secured by collective work effort. Similar to other AFNs like Fair Trade, this domestic food economy is partially separated from capitalist market networks, and the resulting material circuits of food (and privilege) are reproduced through moral actions and justifications that variably connect people from within and without this alternative space. Unlike most AFNs, however, Cuba's domestic moral economy of food is unified by a long-term revolutionary project that links all Cubans, in one way or another, to a collective value system.

Though influenced by global market values like consumer preferences and competitive prices, I have argued that Tutaños uphold nationalist and socialist values related to historical ideas of justice and civic responsibility in Cuba, particularly the emphasis on collective needs over individual desires. Thus as we can see in chapter 3, Tutaño consumers associate being Cuban with being a 'fighter' who continues to resist adversities such as food insecurities that partly result from centralized re-distribution and monetary controls. The idea of re-distributive justice in Cuba, as instituted through policies like Raul Castro's recent economic reforms, is linked to an ideal reciprocal contract between citizens as workers and the welfare state, a relationship addressed in chapter 4. In chapter 5, I show how this ideal exchange relationship influences how many Tutaños conceptualized and practised 'just' forms of exchange in everyday life. At least during my period of fieldwork, self-interested exchanges to satisfy personal or household needs or desires were tempered by social values such as the idea of revolutionary 'culture' and instituted through re-distributive economic institutions such as the Acopio. Similarly, chapter 6 revealed how small-scale producers of foodstuffs in Cuba are driven by socio-cultural, environmental and economic ideas of justice and responsibility such as the value of hard work for the collective patrimony (e.g. land and its products), and by particular forms of governance such as the proliferation of agroecological experts.

Taken together, the ethnographic material presented in chapters 3–6 points to normative and practical connections between food consumption, exchange and production in Cuba, provisioning processes shaped by an overarching ideal of national sovereignty. In this way, the domestic food economy in Cuba resembles other AFNs, most notably the *Via Campesina* (Farmer Road) Movement, which, as described in chapter 6, is a global network of small farmers who initiated the first transnational project for food sovereignty. The concept of food sovereignty marks food as a special object that cannot (always) be commoditized, and food becomes a trope through which projects for alternatives become linked to wider political, cultural and moral objectives. This characterization of food sovereignty typifies AFNs more generally, for as Susanne Freidberg argues: 'Food has become a flash point of many different struggles in defence of particular cultural norms as well as cultural sovereignty more generally'(Freidberg 2003: 6).

The Cuban project for agroecology and food sovereignty is an important case study of this phenomenon. For AFNs in Cuba are tied to a cultural idea of sovereignty which, having emerged through particular historical and geographical relations, in turn establishes another idealized history and geography separated from neoliberal networks. As I argue in chapters 1 and 2, Cuban nationalism is characterized by what Kapcia (2000) calls an 'ideology of dissent' or '*cubania rebelde*' (rebel Cuban-ness), a history that largely emerged from Cuba's relation to the contrasting economic logics and practices of the US 'Goliath'. In light of this history and geography, it is perhaps not surprising that the revolutionary government of post-1959 Cuba managed to create a relatively isolated 'sovereign' sphere of public goods, with more or less success depending on one's point of view.

Critical geographers of alterity question projects that centre on one particular scale (e.g. the 'local') for establishing a 'moral hierarchy' (Jonas 2010: 20) that may not actually contribute to positive change (see for example, studies of the social economy and devolution in the UK, e.g. Amin *et al.* 2003). By emphasizing the scalar politics of Cuban food sovereignty, however, I have not moralized the subject except in the terms that the culture itself adopts. My choice of emphasis has not been to romanticize the nation in Cuba, nor to idealize Cuban socialism, rather to reveal how ordinary people in Tuta make sense of their own contradictory economic worlds. In relating 'is' to 'ought' from both the top-down and the bottom-up, I have shown how Cuba's alternative project for community is 'contested, recognised or not recognised, in various ways through ... discussion of ... food provisioning practices'(Holloway *et al.* 2010: 162).

Alternative systems of food provisioning in Cuba: potentialities and drawbacks

Given the hierarchical structure of the Cuban Communist Party (as discussed in chapters 4 and 5), it has been useful to frame my analysis in terms of the systems of provisioning perspective, for the latter emphasizes vertical relations between particular consumption, exchange and production paths rather than horizontal relations within any one of these provisioning 'moments'. The systems of provisioning approach has allowed me to fit my theoretical analysis with the empirical realities of Cuban food politics, for the domestic food economy in Cuba is controlled by hierarchical institutions such as the Acopio, which create an enclave around food designated for those in need. Ben Fine and Ellen Leopold's (1993; see also Fine 2002) concept of 'provisioning' – vertical connections between production and consumption – is also relevant to the cultural material, for in Cuba food provisioning is underpinned by a variant of nationalism in which values such as self-sacrifice, hard work and asceticism connect moralities of production to moralities of consumption. As Ben Fine (2002, cited by Mansvelt 2005: 110) argues 'a cultural system is attached to each system of provision'.

Restricting flows of provisioning that link food consumption, exchange and production in Cuba leads to inequalities which, as I have argued, are often justified in terms of the national moral economy (I return to the issue of inequalities below). But vertical flows between such economic 'moments' in the food chain may also be more conducive to alternative food networks than horizontal flows, for as Watts *et al.* argue:

> The terms "vertical" and "horizontal" are open to critique, [but] this does not mean that the ideas underlying them are unsound. The conceptual differentiation between networks that operate within a defined area and those that link it to others, and the greater potential of the former to generate endogenous economic development, are relevant to alternative systems of food provision. (Watts *et al.* 2005: 31)

Though limited by macroeconomic constraints, restricted flows of food in Cuba are directed and redirected to feed localized or national communities. Domestic food provisioning in Cuba is thus characterized by a kind of 'endogenous economic development' made possible through power geometries of the state but also through symbolic 'convergences' (Needham 1975) between top-down and bottom-up values. Socially and environmentally beneficial provisioning practices such as reciprocal exchange networks and agroecology are justified and actualized in terms of a national value system that becomes 'personalized' (Cohen 2000: 150) as it enters into everyday life:

> People construct the nation through the medium of their own experience, and in ways which are heavily influenced by their own circumstances. The nation is mediated through the self. ... [P]eople claim the nation for themselves by claiming it *as* themselves, and thereby make its interests and theirs ... coincident. (Cohen 2000: 146)

Thus as we have seen, agroecological farmers in Cuba take pride in producing sustainably for the nation and for their own communities, traders who hinder food access are morally condoned by officials and ordinary Tutaños alike, and even consumers – those most restricted by the cultural politics of Cuban socialism – often uphold national values of resistance and self-sacrifice in their rationalizations of food scarcities and uneven access.

As in (post)socialist countries,[2] food provisioning practices in Cuba are tied to socialist subjectivities. But what makes domestic food provisioning in Cuba so unique is the durability of its networks, a resilience that stems from its long-term project for national sovereignty. Cuban citizens have been intimately related to this project; indeed, as Kapcia (2008: 44) argues, the endurance of the Cuban Revolution is due not to the minority whom he calls 'unquestioningly loyal activists', nor to the other estimated one-fifth to one-third of the population whom he classifies as dissidents. Rather, the staying power of revolutionary spatialities, materialities and moralities have to do with what Kapcia sees as the remaining and critical 1/3rd to 1/2 [of the population], who have tended to remain passively loyal to 'the Revolution', but who have been more than prepared to complain, to operate in the black economy, to ignore calls to mobilize, and so on, those whose loyalty has been focused on nationalism, social benefits, or Fidel Castro's personality, but those who have feared the alternatives. (Kapcia 2008: 44)

I have argued in this book that ordinary economic life in Tuta reflects the spatialities, materialities and moralities of Cuban nationhood. Given stark inequalities of access in Cuba, I have been cautious not to romanticize such connectivities; nor have I privileged the domestic food economy to the exclusion of wider market networks accessible to a select few. As I have found through long-term ethnographic fieldwork, people in Cuba, including myself, inhabit spaces of provisioning variously according to their material and moral positioning vis-à-vis inside and outside economic relations and forms of justification. While I could shift into and out of tourist market spaces, many of my Cuban friends could or would not, for it was unacceptable to be seen in public places reaping material benefits from a 'rich' foreigner.

Such inequalities should give one pause in using terms like 'alternative'. But the *reproduction* of alternative systems of provisioning in Cuba is due not only to political dogmatism and material necessity, nor simply to a kind of 'capitalism of the poor' (Quijano 2001, cited by Williams 2005: 177), a term others have rightly used in criticizing romantic portraits of non-commoditized exchanges among poor people in other places. Non-commoditized relations in Tuta that work within or between commoditized relations may result from political economic power and inequalities, but they are *also* moralized in terms of Cuban nationalism, which in some ways justifies suffering for a higher, transcendental cause. In this sense, AFNs or systems of provisioning in Cuba are successful, for as Holloway *et al.* (2010: 162) argue, 'contingent, accidental and partial commitment[s] do not undermine ... success'.

Geographers of alternative economic spaces are generally sympathetic with the aims of the projects they study, but they also view them critically. Here I have presented the positive aspects of the moral economies of food provisioning in Cuba, but this book has also been a critique of Cuba's cultural politics. Though it emphasizes social embeddedness and sustainability, restricted systems of food provisioning in Cuba also limit potentialities of ordinary economic life. As Andrew Sayer (2001: 698) has argued: 'The metaphor of em*bed*dedness sounds soft and comforting, and possibly sends our critical faculties to sleep, but what it describes can be harsh and oppressive on occasion' (italics in original). Sayer (2011: 154) claims that norms related to 'rights or matters of distributive justice regarding scarce resources' are 'more categorical and strictly enforced'. This is certainly true for the case of Cuba, since the value of food and other public goods as entitlements (rather than commodities) shapes the way the domestic food economy is organized and controlled. By creating a boundary around acceptable and unacceptable forms of provisioning, both national and everyday moral economies of food in Cuba reflect deep social anxieties about the present and future state of Cuban economy and society. They also serve to reproduce what Mary Douglas and Baron Isherwood have called the 'total system of work and reward':

> The energy to protect boundaries from unwanted arbitrage is stoked by the sense of how much is at risk – personal survival in a competitive scene or collective survival in a whole system of values and traditions. Restricted circulation is not likely to flourish except as an element of the total system of work and reward. (Douglas and Isherwood 1978: 141)

Like all projects that emphasize unity, 'idealist visions' (Fuller and Jonas 2003, cited by Holloway *et al.* 2010: 164) such as that for Cuban socialism may hinder some forms of sociality and economic diversity, particularly those that do not fit into the prevailing 'system of work and reward'. Cuba's alternative provisioning system is based on a kind of cultural imperialism that limits 'rewards' to hard workers or those recognized as needy. Its 'alternative' is based on another abstract model of economy that does not always correspond to localized values and circumstances.

As geographers of ethical consumption have warned, 'purist positions' (Renard 2003, cited by Goodman 2004: 903) that denounce all forms of market activity as negative are often no better than neoliberal projections for economic change, for both rely on moralized notions of 'ought' instead of real-life experiences and practices of economy. For instance, abstract moral economies of alterity often reject 'the consumer' as an egoistic subject. To use the terms of Raul Castro, he or she is a 'parasite' that takes from rather than contributes to the civic community:

> Proponents of the market and consumer choice think that people *should* act like this, despite lots of evidence that they *don't*. Critics of the market tend to assume that people *do* act like this, but they think that they *ought not to*, and therefore intone them to act more responsibly. For critics of consumerism and consumerization, the shared characterization of 'the consumer' as individualized and egoistic allows a rather idealized model of 'citizenship' to be counter-posed to the apparent rise of consumerism. In this model, being a citizen is all about the selfless pursuit of the common good undertaken in collective cooperation with others. (Barnett *et al.* 2011: 29; italics in original)

Similar to some political and academic critiques of ethical consumption as 'neoliberalizing' in its focus on consumers, the domestic system of food provisioning in Cuba is underpinned by an 'exclusionary model' (Lee 2010: 282) that upholds the western binary between market and mutuality. As in all places, however, market and non-market spaces and logics in Cuba interrelate in everyday life.

Figure 7.2 A couple relax on their small balcony. Source: author.

Assumptions about Cuban transition – whether socialist or liberal – limit the potentialities of both spatial and economic life in Cuba. As Anna Cristina Pertierra argues, an idea of Cuban transition that

> unquestioningly reproduces a capitalist fantasy that all Cuban consumers crave Nike and McDonald's and dream of nothing more than moving to Miami, is as pointless to any genuine understanding of Cuban society as [is] a socialist fantasy that the revolution's provision of hospitals and schools has somehow erased all Cubans' desires for plentiful consumer commodities. (Pertierra 2007: 46–7)[3]

Whether more closely tied to global market circuits or less, market activity coexists (and sometimes even supports) non-market activity in Cuba. Thus while the ideas and practices of ordinary Cubans are embedded in the moral and political economy of the nation state, the same Cubans may also create moral and material networks that extend beyond national borders.

Towards value pluralism[4]

In a recent critique of John Rawls's (1971) idea of justice, Amartya Sen uses as an analogy the differentiation in classical Sanskrit between *niti* (jurisprudence) and *nyaya* (ethics). He compares the distinction to that between what he calls 'arrangement-focused' and 'realization-focused' ideas of justice, respectively (Sen 2009: 20). While *niti* means 'organizational propriety and behavioural correctness', *nyaya* means 'realized justice'. Sen criticizes Rawls's version of justice as 'transcendental institutionalism' because it is concerned with a definition of *the* perfect justice (based on a hypothetical model of humans in a pre-class state who would favour re-distribution to the least well-off), rather than a variety of reasons for justice that emerge from actual experiences and relations (Sen 2009: 5–6). Instead of favouring one idea of justice (e.g. re-distribution) over others, Sen argues for an idea of justice that takes into account different positionalities, any one of which may be relevant according to the event in question.

> A feminist activist in America who wants to do something to remedy particular features of women's disadvantage in, say, [South] Sudan would tend to draw on a sense of affinity that need not work through the sympathies of the American nation for the predicament of the [South] Sudanese nation. (Sen 2009: 137)

Similar to Sen's critique of singular notions of justice, geographers have argued that dominant theories like economic globalization foreclose a multiplicity of relations people make in everyday life (Gibson-Graham 1996, 2006; Lee 2006; Massey 2011 [2005]). In Sen's terms, neoliberal theories depend on a uniform concept of justice that hinges on the market as *the* transcendental institution. Counter-theories that oppose the impersonal nature of markets are often no better at recognizing heterogeneous positionalities, for they may substitute another kind of 'transcendental institutionalism' for the liberal market. Thus, as Clive Barnett *et al.* (2011: 12) argue, and as referred to above, people who criticize 'shopping for change' as inauthentic or apolitical overlook the kind of 'realized justice' a politicized form of market consumerism may actually engender. Like other forms of 'responsible' action, such accounts of ethical consumption do not 'narrow the gap between ascription and practice' (Noxolo *et al.* 2012: 424). In Sen's terms, the *niti* of market-based campaigning does not always coincide with the *nyaya* of everyday life. When, however, ethical campaigns *do* take into account the way people identify themselves – as self-interested consumers *as well as* responsible agents for change – then a broader, empirically based understanding of social change may develop (Barnett *et al.* 2011). Such 'post-moralistic' (Schudson 2006, cited by Barnett *et al.* 2011: 3) approaches to consumption and other provisioning processes open up possibilities for the creation of multi-scalar socio-economic and political relationships based on various identifications and capabilities.

The concept of 'capabilities' – perhaps the most famous aspect of Sen's work (see Sen 1999; Nussbaum 2000) – is essential for his more recent appeal to an idea of justice that takes 'actually-existing societies' into account (Sen 2009: 83, 85). Sen defines capabilities as the opportunity and/ or capacity 'to do things [one] has reason to value' (Sen 2009: 231). Following this line of thought, a lack of capabilities is the institutionalized exclusion of certain values, such as the needs of disabled people, that go beyond welfare economics (Sen 2009: 203). Sen's defence of the capability approach does not extend to a relativist position that sees all values as equally beneficial to the present and future of humanity, however. Adopting wholeheartedly the Cuban value system, for example, may hinder rather than help the capabilities of people living in Cuba, whose everyday struggle to work for a decent wage and to find affordable food would not be as challenging if the Cuban state did not control domestic markets so strictly.

While value pluralism furthers justice in the sense that it takes into account the 'equal moral worth of all human beings' (Fraser 2009), taking a relativist stance that does not differentiate between 'good' and 'bad' forms of value may hinder rather than help projects for social, economic and environmental justice. In order to pursue a kind of justice that is inclusive, if limited to the achievable, it is therefore essential to create the conditions for public debate, which, through reasoned argument, may eliminate what Sen calls 'positional illusions' (Sen 2009: 170):

> Consider a person who belongs to a community that does not have familiarity with distant-dependent projections, nor with any other source of information about the sun and the moon. Lacking the relevant conceptual frameworks and ancillary knowledge, that person may decide, on the basis of positional observations, that the sun and the moon are indeed of the same size ... His belief ... is, of course, a mistake (an illusion), but [it] cannot, under the circumstances, be seen as purely subjective, given the totality of his positional features. Indeed, anyone in exactly his position ... can understandably take much the same view, prior to critical scrutiny, for much the same reasons. (Sen 2009: 168)

Sen goes on to discuss women in some places of the world (e.g. India) who may justify unequal treatment in public and private spaces by appealing to cultural norms. As Sen explains, due to positional illusions (such as community pressures and practices of shaming) such a woman may actually exacerbate her own subordination through self-exploiting relations of care in the household (Sen 2009: 166–69). I have argued in this book that inequalities in Cuba resulting from institutional and moral separations between market and non-market spaces are also justified in terms of a particular cultural politics, which, among other things, affects the capabilities of people who value consumer goods at least as much as the Cuban Revolution.

In order to ensure that value pluralism assists rather than hinders justice, we must promote and, if possible, institute values that amplify the shared 'goods' of humanity – gender, sexuality, class and racial equality (in organizational as well as habitual contexts); clean air, land and seas; healthy and culturally appropriate food that is accessible and affordable (and, as much as possible, grown close to home); universal access to public spaces and transportation, and so on – and discourage values that infringe upon the capabilities of present and future peoples (or indeed one's own capabilities). Knowledge of 'thick' notions of justice that emerge in place, such as relations of responsibility and care in Cuba, must be combined with statements about universal or 'thin' requirements for justice (Smith 2000), such as the need to ensure the social, economic and ecological well-being of this generation and those that follow. A combined approach that incorporates both abstract and experiential knowledge opens up possibilities for economic agents in places like Cuba, where the economic capabilities of people are presently restricted by the militant particularism of the Cuban nation state. A plural approach to justice would recognize the importance of non-market, place-based values in Cuba like solidarity and sustainability, but also of market values such as consumer preferences and affordability.

Taking 'thick' as well as 'thin' ideas of justice into account, the researcher or policymaker may discover certain combinations of market and non-market values that are better for the present and future of humans and non-humans than others. For instance, 'good' commodification may promote economic growth while limiting its definition to environmentally benign activity, as recently suggested by the *People and the Planet* report of the Royal Society (2012). Cuba's admirable strategy to support agroecological farming may also be improved by eliminating present restrictions on farmer-to-farmer seed exchanges, unreasonable controls over the marketing of farmers' produce or the threat of a prison sentence for slaughtering one's own cattle. There are many worthy models that may be utilized in the pursuit of justice; what is needed is the flexibility to ensure that such models are not mired by the positional illusions of distant model-makers.

Sen's solution to the problems of positional illusions draws from Adam Smith, who argued for the need to enlist an impartial spectator – someone who is distant both geographically and culturally – when making arguments about how to improve society (Sen 2009: 108, 407). Smith's call for an impartial spectator meant to avoid the militant particularism of place-based models, opening to public discussion different perspectives based on different forms of value, even if agreement is sometimes impossible.

To listen to distant voices, which is part of Adam Smith's exercise of invoking the 'impartial spectator', does not require us to be respectful of every argument that may come from abroad. ... We may reject a great many of the proposed arguments – sometimes even all of them – and yet there would remain particular cases of reasoning that could make us reconsider our own understandings and views, linked with the experiences and conventions entrenched in a country, or in a culture. (Sen 2009: 407)

An understanding of other perspectives allows us to question our own norms and values as well as those of others, and to reason whether cultural justifications, such as the liberal idea that all individuals are 'equally capable of looking after themselves' (Macphereson 1979 [1962]: 245), are really beneficial to *all* humans and living things.

Methodologically, value pluralism carries both a risk and a responsibility. For those who have the power to create knowledge about 'others' through representation, it is essential to emphasize the provisional and testable nature of this knowledge, to be as flexible as society itself in constantly revising what is known to fit real-life experiences (and not the other way around). While there is always the risk of reification, the ethnographic researcher has the responsibility to open public understanding to particular values that have formed over time in the (culturally if not always geographically established) 'place' in question. In the process, one must acknowledge his or her own situated values, but also position oneself so as to know whether or not any capabilities of the people under study are limited by the normative modalities of Leviathans. This means moving beyond an idea of justice based on transcendental models and institutions – like socialism – to a project for justice that takes into account how such ideals and their organizational manifestations actually work in practice. It is no good 'going native', for this does not permit the kind of comparative understanding of capabilities enabled when the researcher attempts to view the world from the eyes of Adam Smith's impartial spectator.

The comparative, non-parochial study of societies and economies, a central theoretical method in social anthropology, should be of use to geographers interested in issues of justice, value and shifting partialities. Cultural relations between material needs and moral demands set the standards of what it means to be human within each place, for as Marshall Sahlins (1976: 168) argues, humans 'do not merely "survive". They survive in a definite way'. Long-term ethnographic fieldwork is indeed essential, since 'the question of what it is to be human is not merely of philosophical interest. It has profound socio-material significance too. It should, therefore, continue to inform a critical social science' (Lee 2010: 275). This book provides a means to such ends by revealing both positive and negative aspects of an alternative food provisioning network shaped and reproduced by cultural values that have emerged in time and space. By experiencing and writing about another human world, I have made it possible to consider the material needs and moral demands of a particular population of distant 'others'.

Figure 7.3   Source: author.

Notes

1   This aligns with what Ben Page (2005: 294, 303, in reference to Castree 2001, 2004) calls the 'new geography of commodities', which not only accounts for multiple and diverse relations of value but offers normative grounds to define 'good' and 'bad' forms of commodification.
2   See, for instance, Klein (2008, 2009) for China; Caldwell (2009) for various countries in the post-Soviet world; Stenning *et al.* (2010: 71, 144–55) for Poland and Slovakia; Caldwell (2011) for Russia.
3   This quote is also cited in Wilson (2012: 287).
4   Fraser (2009).

References

Amin, Ash, Angus Cameron and Ray Hudson. (2003) The alterity of the social economy. In *Alternative Economic Spaces*, edited by Andrew Leyshon, Roger Lee and Colin C. Williams. London: Sage Publications, pp. 26–54.

Barnett, Clive, Paul Cloke, Nick Clarke and Alice Malpass (eds). (2011) *Globalizing Responsibility: the Political Rationalities of Ethical Consumption*. Oxford: Wiley Blackwell.

Bohannan, Paul and George Dalton. (1962) Introduction. In *Markets in Africa*, edited by Paul Bohannan and George Dalton. Evanston: Northwestern University Press, pp. 1–26.

Caldwell, Melissa (ed.). (2009) *Food and Everyday Life in the Post-Socialist World*. Bloomington: Indiana University Press.

Caldwell, Melissa. (2011) *Dacha Idylls: Living Organically in Russia's Countryside*. Berkeley: University of California Press.

Callon, Michel and Bruno Latour. (1981) Unscrewing the big leviathan: how actors macro-structure reality and how sociologists help them to do so. In *Advances in Social Theory and Methodology: Towards an Integration of Micro and Macro-Sociology*, edited by K. Knorr-Cetina and A.V. Cicouvel. Boston: Routledge, pp. 277–303.

Castree, Noel. (2009) The time-space of capitalism, *Time and Society* 18(1): 26–61.

Cohen, Anthony P. (2000) Peripheral vision: nationalism, national identity and the objective correlative in Scotland. In *Signifying Identities: Anthropological Perspectives on Boundaries and Contested Values*, edited by Anthony P. Cohen. London: Routledge, pp. 145–69.

Douglas, Mary and Baron Isherwood. (1978) *The World of Goods: Towards an Anthropology of Consumption*. London: Allen Lane.

Dumont, Louis. (1977) *From Mandeville to Marx: the Genesis and Triumph of Economic Ideology*. Chicago: University of Chicago Press.

Fine, Ben. (2002) *The World of Consumption. The Material and Cultural Revisited*. London: Routledge.

Fine, Ben and Ellen Leopold. (1993) *The World of Consumption*. London: Routledge.

Fraser, Nancy. (2009) *Scales of Justice: Reimagining Political Space in a Globalizing World*. New York: Colombia University Press.

Freidberg, Susanne. (2003) Not all sweetness and light: new cultural geographies of food. *Social and Cultural Geography* 4(1): 3–6.

Fuller, Duncan and Andrew E.G. Jonas. (2003) Alternative financial spaces. In *Alternative Economic Spaces*, edited by Andrew Leyshon, Roger Lee and Colin Williams. London: Sage Publications, pp. 55–73.

Fuller, Duncan, Andrew E. G. Jonas and Roger Lee. (2010) *Interrogating Alterity: Alternative Economic and Political Spaces*. London: Ashgate.

Gibson-Graham, J.K. (1996) *The End of Capitalism (as We Knew it): a Feminist Critique of Political Economy*. Minneapolis: University of Minnesota Press.

Gibson-Graham, J.K. (2006) *A Postcapitalist Politics*. Minneapolis: University of Minnesota Press.

Giddens, Anthony. (1984) *The Constitution of Society. Outline of the Theory of Structuration*. Cambridge: Polity Press.

Goodman, Michael K. (2004) Reading fair trade: political ecological imaginary and the moral economy of fair trade foods, *Political Geography* 23: 891–915.

Hodkinson, Stuart (2010) Housing in common: in search of a strategy for housing alterity for England in the 21st century. In *Interrogating Alterity: Alternative Economic and Political Spaces*, edited by Duncan Fuller, Andrew Jonas and Roger Lee. London: Ashgate, pp. 241–258.

Holloway, Lewis, Rosie Cox, Moya Kneafsey, Elizabeth Dowler, Laura Venn and Helena Tuomainen. (2010) 'Are you alternative?' 'Alternative' food networks and consumers' definitions of alterity. In *Interrogating Alterity: Alternative Economic and Political Spaces*, edited by Duncan Fuller, Andrew Jonas and Roger Lee. London: Ashgate, pp. 161–74.

Jackson, Peter, Neil Ward and Polly Russell. (2008) Moral economies of food and geographies of responsibility, *Transactions of the Institute of British Geographers* (N.S.) 34: 12–24.

Jonas, Andrew. (2010) 'Alternative' this, 'alternative' that … : Interrogating alterity and diversity. In *Interrogating Alterity: Alternative Economic and Political Spaces*, edited by Duncan Fuller, Andrew Jonas and Roger Lee. London: Ashgate, pp. 3–30.

Kapcia, Antoni. (2000) *Cuba: Island of Dreams*. Oxford: Berg.

Kapcia, Antoni. (2008) *Cuba in Revolution: a History since the Fifties*. London: Reaktion Books.

Klein, Jakob (2008) Afterword: comparing vegetarianisms, *South Asia Journal of South Asian Studies* 31(1): 199–212.

Klein, Jakob. (2009) Creating ethical food consumers? Promoting organic foods in urban Southwest China, *Social Anthropology* 17(1): 74–89.

Lee, Roger. (2006) The ordinary economy: tangled up in values and geography, *Transactions of the Institute of British Geographers* (N.S.) 31: 413–32.

Lee, Roger. (2010) Spiders, bees or architects? Imagination and the radical immanence of alternatives/diversity for political-economic geographies. In *Interrogating Alterity: Alternative Economic and Political Spaces*, edited by Duncan Fuller, Andrew Jonas and Roger Lee. Surrey: Ashgate, pp. 273–88.

Lee, Roger. (2011a) Ordinary economic geographies: can economic geographies be non-economic? In *Handbook of Economic Geography*, edited by Andrew Leyshon, Roger Lee, Linda McDowell and Peter Sunley. London: Sage Publications, pp. 368–82.

Lee, Roger. (2011b) Within and outwith/material and political? Local economic development and the spatialities of economic geographies. In *Handbook of Local and Regional Development*, edited by Andy Pike, Andrés Rodríguez-Pose and John Tomaney. London: Routledge, pp. 193–211.

Leyshon, Andrew, Roger Lee and Colin C. Williams (eds). (2003) Introduction. In *Alternative Economic Spaces*, edited by Andrew Leyshon, Roger Lee and Colin C. Williams. London: Sage Publications, pp. 1–26.

Macpherson, C.B. (1979 [1962]) *The Political Theory of Possessive Individualism. Hobbes to Locke*. Oxford: Oxford University Press.

Mansvelt, Juliana. (2005) *Geographies of Consumption*. London: Sage Publications.

Massey, Doreen. (2011 [2005]) *For Space*. London: Sage Publications.

Miller, Daniel. (1997) *Capitalism: an Ethnographic Approach*. Oxford: Berg.

Needham, Rodney. (1975) Polythetic classification: convergence and consequences, *Man* (N.S.) 110: 349–69.

Noxolo, Pat, Parvati Raghuram and Clare Madge. (2012) Unsettling responsibility: postcolonial interventions, *Transactions of the Institute of British Geographers* (N.S.) 37(3): 418–29.

Nussbaum, Martha C. (2000) *Women and Human Development: the Capabilities Approach*. Cambridge: Cambridge University Press.

Page, Ben. (2005) Paying for water and the geography of commodities, *Transactions of the Institute of British Geographers* (N.S.) 30: 293–306.

Pertierra, Anna Cristina. (2007) Cuba: the struggle for consumption, doctoral thesis, University College London.

Rawls, John (1971) *A Theory of Justice*. Cambridge: Belknap Press of Harvard University Press.

Royal Society (2012) *People and the Planet*, final report, available at http://royalsociety.org/policy/projects/people-planet/report/ (last accessed 16 May 2013).

Sahlins, Marshall. (1976) *Culture and Practical Reason*. Chicago: University of Chicago Press.

Sayer, Andrew. (2000) Moral economy and political economy, *Studies in Political Economy* 61: 79–104.

Sayer, Andrew. (2001) For a critical cultural political economy, *Antipode* 33(4): 687–708.

Sayer, Andrew. (2011) *Why Things Matter to People: Social Science, Values and Ethical Life*. Cambridge: Cambridge University Press.

Sen, Amartya. (1999) *Development as Freedom*. New York: Knopf.

Sen, Amartya. (2009) *The Idea of Justice*. London: Penguin Books.

Smith, David. (2000) *Moral Geographies: Ethics in a World of Difference*. Edinburgh: Edinburgh University Press.

Stenning, Alison, Adrian Smith, Alena Rochovská and Dariusz Świątek. (2010) *Domesticating Neo-liberalism: Spaces of Economic Practice and Social Reproduction in Post-Socialist Societies*. Oxford: Wiley Blackwell.

Watts, D.C.H., B. Ilbery and D. Maye. (2005) Making reconnections in agro-food geography: alternative systems of food provision, *Progress in Human Geography* 29(1): 22–40.

Williams, Colin C. (2005) *A Commodified World? Mapping the Limits of Capitalism*. London: Zed Books.

Wilson, Marisa. (2012) The moral economy of food provisioning in Cuba, *Food, Culture and Society* 15(2): 277–91.

Appendix 1
Key Political Economic Events of the Cuban Revolution

Timeline of the key political economic events of the Cuban Revolution, 1959–2013.

1959	Fidel Castro, Ernesto 'Che' Guevara, Raul Castro and Camilo Cienfuegos march triumphant into Havana; former President Batista flees the country; Fidel Castro becomes Prime Minister of Cuba
1960	US businesses in Cuba are nationalized
1961	Fidel Castro declares the Revolution socialist
United States breaks diplomatic ties with Cuba
United States sends Cuban exiles to Cuba's Bay of Pigs, who are defeated
As head of the National Bank, Guevara replaces all pre-Revolutionary Cuban pesos with Revolutionary pesos
1962	Cuban missile crisis
Organization of American States suspends Cuba's membership
1963–1965	The Great Debate
1965	The Cuban Communist Party becomes the sole political party in Cuba
1968	Castro launches the 'Revolutionary Offensive' campaign nationalizing most Cuban businesses
1969	Fidel Castro launches the 'Ten Million Tons of Sugar' campaign for 1970

Everyday Moral Economies: Food, Politics and Scale in Cuba, First Edition. Marisa Wilson.
© 2014 John Wiley & Sons, Ltd. Published 2014 by John Wiley & Sons, Ltd.

1972	Cuba joins the USSR's Council for Mutual Economic Assistance
1975	First Congress of the Cuban Communist Party
1976	Fidel Castro becomes president
	The first Constitution of the Cuban Republic under the Revolution is ratified
1980	Second Congress of the Cuban Communist Party
1986	Third Congress of the Cuban Communist Party
	The 'Rectification' campaign is launched
1989	Fall of the Berlin Wall
1991	Break up of the Soviet Union
	Fourth Congress of the Cuban Communist Party
	Fidel Castro announces the period as a 'Special Period in Time of Peace'
1992	The Second Constitution of the Republic of Cuba under the Revolution is ratified
1993–1995	Most acute shortages of the 'Special Period in Time of Peace'
1992	United States enacts the Torricelli Act, preventing food and medicine from entering Cuba (with the exception of humanitarian aid)
1993	The Cuban government legalizes hard currency (US dollars) in the state sector; other economic openings begin
1997	Fifth Congress of the Cuban Communist Party
1996	United States enacts the Helms–Burton Act, penalizing foreign companies trading with Cuba
1999	The 'Battle of Ideas' campaign is launched
2000	United States enacts the Trade Sanctions Reform Act, which permits the export of alimentary and medical commodities to Cuba, but prohibits US financing of such exports
2002	Amendments to the 1992 Constitution are ratified
2004	The Cuban government legalizes hard currency for domestic transactions, converting the US dollar to the Cuban Convertible Dollar
2006	Fidel Castro is hospitalized
	The year is hailed the 'Year of the Energetic Revolution'
2008	Raul Castro becomes president
2011	Economic reforms are passed in Cuba that allow for the purchase and sale of houses, cars and a wider opening of the economy to private forms of work
	Sixth Congress of the Cuban Communist Party
2013	Cuban travel to and from Cuba is legalized (with restrictions)

Appendix 2
Daily Nutritional Requirements in Cuba

Figure showing the daily per capita intake of calories, proteins and fat, 1980–1999. MR/day (triangles): the Food and Agriculture Organization's (FAO's) minimum daily nutritional requirements for Cuba according to age, height and weight – 2400 kcal of calories, 75 g of fat and 72 g (of which 29 g and 43 g of animal and vegetable origin, respectively) of proteins (MR, minimum requirement). Official (squares): official statistics in Cuba's *Anuario Estadístico*. FAO (dots): FAO's estimates. Source: Álvarez 2004: 6; The issue of food security in Cuba (Paper FE483). Institute of Food and Agriculture Sciences (UF/IFAS), University of Florida, http://edis.ifas.ufl.edu (last accessed 24 May 2013).

Everyday Moral Economies: Food, Politics and Scale in Cuba, First Edition. Marisa Wilson.
© 2014 John Wiley & Sons, Ltd. Published 2014 by John Wiley & Sons, Ltd.

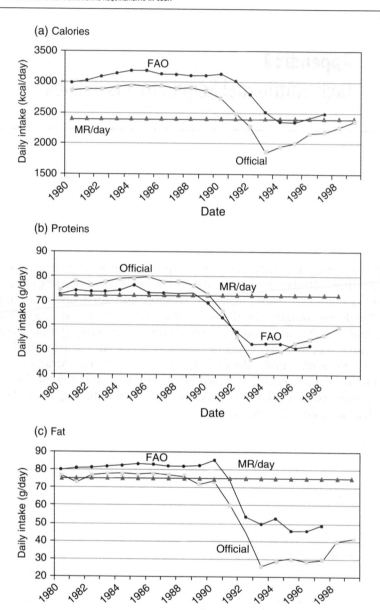

(a) Calories

(b) Proteins

(c) Fat

Appendix 3
Institutional Levels for National Food Provisioning

Flow chart showing the institutional levels for national food provisioning in Cuba.

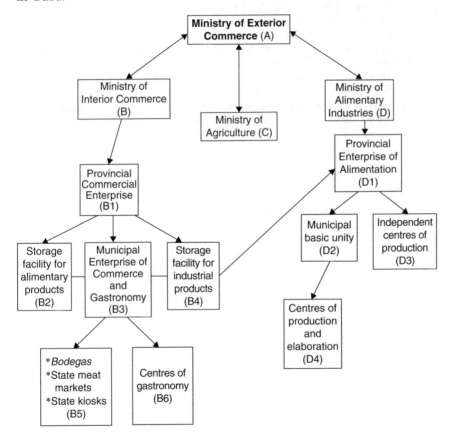

Everyday Moral Economies: Food, Politics and Scale in Cuba, First Edition. Marisa Wilson.
© 2014 John Wiley & Sons, Ltd. Published 2014 by John Wiley & Sons, Ltd.

A: The Ministry of Exterior Commerce (A) receives imports of foodstuffs and food products from the three lower ministries: B, C and D.

B: With information from the Provincial Commercial Enterprise (B1) regarding the alimentary necessities of each province, the Ministry of Internal Commerce (B) distributes foodstuffs to B1.

B1: The Provincial Commercial Enterprise (B1) distributes foodstuffs to B2, B3 and B4 after receiving information from the Municipal Enterprise of Commerce and Gastronomy (B3) regarding the alimentary necessities of the corresponding municipality.

B3: After consulting local institutions such as local Ministries of Public Health, which record births and deaths, and local Registry Offices for Consumers, which distribute *libretas* or ration cards, the Municipal Enterprise of Commerce and Gastronomy (B3) distributes foodstuffs to B5 and B6. B6 are state restaurants or other small centres where cooked food is purchased.

C: The Ministry of Agriculture (MINAG, C) works much like the Ministry of Exterior Commerce described here.

D: The Ministry of Alimentary Industries (D) receives raw materials from A, if these cannot be acquired in Cuba. If the raw materials are produced in Cuba, D may receive foodstuffs from either B2 or B4.

D2, D3: Receive food products from D1. D2 consists of centres classified into categories, such as the *panaderia* (bread store).

D3, D4: consists of state factories, where products such as spaghetti are manufactured; D4 are work centres that receive food from D2.

Appendix 4
Monthly Food Rations per Person

Table showing the approximate monthly food rations per person in Tuta in 2011.

Product	Quantity	Price (in Cuban pesos)
Rice	5 lb/2.3 kg	0.70/lb (1.4/kg)
Beans (black, red, white)	1 lb/0.45 kg	0.32/lb (0.64/kg)
White (refined) sugar	3 lb/1.36 kg	0.15/lb (0.30/kg)
Dark (unrefined) sugar	2 lb/0.9 kg	0.10/lb (0.20/kg)
Milk (only children under 7 years)	1 L/day	0.25/L
Eggs	5 lb/2.3 kg	0.15 each
Potatoes/bananas or other tubers	5 lb/2.3 kg	0.40/lb (0.80/kg)
Bread rolls	1 roll/day	0.05/roll

Everyday Moral Economies: Food, Politics and Scale in Cuba, First Edition. Marisa Wilson.
© 2014 John Wiley & Sons, Ltd. Published 2014 by John Wiley & Sons, Ltd.

Appendix 5
Weekly Household Food Purchases

Table showing the household food purchases over a 1-week period in 2011. The household consists of the mother and father, aged 33 and 37 years, respectively, and the daughter and son, aged 12 and 8 years, respectively. The mother works for the Egyptian Mosquito Brigade (a state programme to inform households about how to avoid mosquito infestation); the father works at a *bodega* (neighbourhood distribution centre for rations); the two children attend school. Only earnings from state work are taken into account.

Food item	Quantity	Point of purchase	Purchaser	Legal or illegal?	Price (in Cuban pesos)
Taro	5 lb	Farmers' market	Mother	Legal	20
Pork meat/fat	2 lb	Neighbour	Mother	Illegal	40
Eggs	5 eggs	Neighbourhood rationing centre	Father	Legal	0.75
Small sweet peppers	0.10 lb	State market	Mother	Legal	5
Pumpkin	0.5 lb	Farmers' market	Mother	Legal	5
Onions	1.5 lb	Farmers' market	Mother	Legal	20

(continued)

Everyday Moral Economies: Food, Politics and Scale in Cuba, First Edition. Marisa Wilson.
© 2014 John Wiley & Sons, Ltd. Published 2014 by John Wiley & Sons, Ltd.

Food item	Quantity	Point of purchase	Purchaser	Legal or illegal?	Price (in Cuban pesos)
Plantains	3 lb	'Grabbed' from work	Father	Illegal	0
Bread	28 rolls	Neighbourhood rationing centre	Father	Legal	1.40
Milk	3 L	Street vendor	Mother	Illegal	15
Rice	3 lb	Neighbourhood rationing centre + 'grabbed'	Mother + father	Legal and illegal	2.10
Spaghetti	0.25 lb	Street vendor	Mother	Illegal	10
Tomato puree	1 bottle (approx. 0.5 L)	State kiosk	Mother	Legal	5
Pig lard	0.5 lb	Neighbour	Mother	Illegal	10

- *Total food expenditure:*
  - 152.75 pesos (equivalent to about 6 CUCs).
- *Other household expenditure:*
  - Children's allowance (mostly for sweets, snacks): 20 pesos.
  - Payments for state credits (for electric kitchenware): 17.5 pesos.
  - Soft drinks: 20 pesos.
  - Electricity bill (monthly): 48 pesos.
- *Total weekly expenditure:*
  - 258.25 pesos.
- *Sum total of weekly earnings (mother + father, state work only):*
  - Approximately 158.75 pesos.

Appendix 6
The Cuban Urban Agriculture Programme

The 29 subprogrammes for the urban agriculture (UA) programme are here paraphrased and translated from: MINAG (Ministerio de la Agricultura, Cuba). (1999) *Lineamentos para los Subprogramas de la Agricultura Urbana para el Año 2000.* Havana: MINAG.

1. Land use:
   a. UA is within a 3-km periphery of all towns and cities.
   b. Crop rotation practised according to climatic, soil, topographic and social conditions.
2. Organic material:
   a. 893 000 m³ of organic material processed per year; applications of 10 kg per m² on state organic farms (*organopónicos*).
   b. Municipal centres process organic matter.
   c. On-farm worm humus production.
3. Seed production:
   d. Use of authorized seeds in each municipality (e.g. beans, cucumbers, pumpkins and other short-cycle crops).
4. Vegetables and fresh herbs:
   a. Produce multiple varieties of crops.
   b. Prioritize leafy vegetables and green beans on *organopónicos*, while in private gardens; increase the cultivation of tomatoes, cabbage, garlic, onions and other plants.
   c. Control the area and production on every farm.
   d. Intercrop according to technical indications.

Everyday Moral Economies: Food, Politics and Scale in Cuba, First Edition. Marisa Wilson.
© 2014 John Wiley & Sons, Ltd. Published 2014 by John Wiley & Sons, Ltd.

5. Popular rice subprogramme:
   a. 150 000 t of rice produced annually with volunteer work on 100 000 ha of state land.
6. Medicinal plants and dried herbs.
7. Ornamental plants and flowers:
   a. Develop and nurture the 'spiritual side of man'.
   b. Produce funerary wreaths.
8. Fruits:
   a. Increase fruit consumption.
9. Forestry and coffee production.
10. Popular plantain/banana production:
   a. Avoid extinction of certain types such as the *platano burro*.
   b. Clone plantains and bananas that are tolerant to black sigatoka disease.
11. Roots and tropical tubers.
12. Comestible oil production.
13. Fodder production.
14. Drainage and irrigation.
15. Apiculture.
16. Aviculture.
17. Sheep and goat rearing.
18. Rabbit rearing.
19. Pig rearing.
20. Aquaculture and fish waste production.
21. Small industry (e.g. tomato purée, guava bars, pig lard).
22. Science, technology and development.
23. Grains.
24. Bovine rearing.
25. Commercialization.
26. Plant protection.
27. Corn and sorghum production.
28. Environmental protection.
29. Greenhouse construction.

Index

Everyday Moral Economies: Food, Politics and Scale in Cuba, First Edition. Marisa Wilson.
© 2014 John Wiley & Sons, Ltd. Published 2014 by John Wiley & Sons, Ltd.

alternatives 24–5, 154–5, 165–6,
    175–6, 181–95
  currencies 165–6, 184–5
  economic geographies and
    provisioning systems 24–5,
    184–90
Altieri, Miguel 154–6, 159
Angola 168
animal traction 158–9, 162
anthropology xiv, xv, xvii, 10–15, 16,
    21, 36–7, 78–9, 84, 87, 90, 100,
    101–17, 134–5, 138, 147–9,
    184–90, 194–5
anti-discrimination policies of Fidel
    Castro 55–6, 63
anti-social behaviours 18
aquaculture 210
Arbelo, Figueroa 160–1, 162
Argentinean origins of Guevara 56
Aristotle 76, 100–1
'arriving' food 105–6
Article 2 of the Rules of the Cuban
    Communist Party (1999) 125–6
Article 7 of Cuba's new labour policy
    102–3
Article 10 of the Rules of the Cuban
    Communist Party (1999) 110
asceticism, culture definitions xviii,
    54–5, 80–1, 110–11, 122,
    186–7
associating prohibitions, tourists
    145–7, 188
'at least now, no one goes to bed
    hungry' 108–9
athletics 90
avocados 76, 83, 139, 170, 173

Baez, Garcia 162
balconies 190
bananas 170, 205, 210
Bantu morals of food 78–9
Barnett, Clive 21, 36, 189, 191–2
Barrios, Adelfo Martin 42, 160, 162
barter trade 10, 35–6, 54, 61–4, 165–6,
    172–3, 184
Basic Units of Cooperative Production
    (UBPCs) 159, 176
basil 170

Batista, President 55, 199
'Battle of Ideas' campaign from 1999
    6–56, 64–5, 106–7, 200
Bay of Pigs incident of 1961 60–1,
    199
beans 1, 6, 76–7, 82, 107, 155–6,
    162–3, 170, 205, 209
beef 81–2, 83–4, 94, 95, 145
beer xi, 18, 62, 101, 130, 147, 148
'before/after idiom' 51, 54–5
Bemba system of food distribution
    101
Berg, Mette Louise 134–5
Berlin Wall fall of 1989 200
Bettleheim, Charles 60, 67
Beveridge, Albert Jeremiah 43
bicycles 165–6
binaries 13–17, 37–66, 91–4, 181–2,
    189–90
biodiversity 155–76
'biological agents', US 171–2
black markets 64, 81–2, 83, 174, 188
Blaine, James G. 43
'blame' model of economic
    responsibility 23, 113–14
blue crabs 83
bodegas 203, 207
Bohannan, Paul 11, 93, 184
Bolívar, Simón 43, 56, 67
Bolivarian Alliance for the Peoples of
    Our America (ALBA) 35
Bolsheviks 134
Boltanski, Luc 34, 73–4
bonuses 103, 113, 125
bottom-up perspectives 146, 149,
    186–95
Bourdieu, Pierre 51, 79
bourgeoisie 38, 42, 47
bovines 210
bread 48–9, 106–7, 143–4, 203, 205,
    207
bribes 137–9
    see also corruption; crime levels
Burke, Timothy 48
Bush, George W. 8, 104

cabbages 209
Callon, Michael xvi, 12–13, 182